# Biotechnology

# BIOTECHNOLOGY
## A NEW INDUSTRIAL REVOLUTION

Steve Prentis

FOREWORD BY

Magnus Pyke

George Braziller, Inc., New York

# Contents

*A photographic section appears between pages 96 and 97*

# Acknowledgments

I should like to thank the many biotechnologists who spared time to discuss their work and its applications. In particular, I am especially grateful to the following for reading and giving me their comments on various sections of the manuscript: Professor Tony Atkinson of the Centre for Applied Microbiology and Research, Dr Norman Cohen of the Open University, Professor Howard Dalton of the University of Warwick and Professor Alan Williamson of Glaxo Group Research. They gave invaluable advice on the presentation of technical information in a way that is, I hope, comprehensible to a wide audience. Any oversimplifications or inaccuracies which remain are, of course, entirely my responsibility. Finally, I should like to express my gratitude to Dr Harmke Kamminga of Birkbeck College, University of London for her continual encouragement and patience in commenting on several drafts of the manuscript.

The author and publisher are also grateful to the following for supplying the plates indicated: Beecham Pharmaceuticals Research Division 4, 5; Biological Photo Service 1; Professor Orio Ciferri 9; Glaxo 6; by courtesy of Kennecott 13; ICI Agricultural Division/John Brown Engineers and Constructors Ltd. 10; Marc Van Montagu 8; Nova 2; Carl E. Shively, Alfred University 12; Stanford University Medical Center 3; V.A.G. (United Kingdom) Ltd. 11; J.W. Watts, John Innes Institute 7.

# Foreword

Technological revolutions, such as those arising from the invention of agriculture, the development of the steam engine, the work of Henry Ford and, now, genetic engineering, have two aspects. The first of these is the technology itself, the second is the intellectual climate which surrounds it.

With regards to the first, biotechnology is, in fact, as old as the hills. Noah knew enough about fermentation to get himself drunk as soon as he landed from the Ark. Leavened bread, cheese, pickled cabbage and cesspools are all examples of biotechnology that go back into antiquity. For many centuries the climate of thought was such that these aspects of applied biology were categorized as useful arts. The age of science could be said to have begun at the end of the eighteenth century with Antoine Laurent Lavoisier's profound statement: 'La vie est une fonction chimique' – life is a chemical process.

Within twenty years of today serious work was being undertaken to improve the fermentation process by which industrial alcohol was produced (mainly from molasses), to isolate organisms capable of producing acetone and butanol or, later on, those more complex molecules, penicillin and streptomycin. What is now being categorized as biotechnology was then part of chemical industry or, at most, fermentation industry. The universities could be depended upon to elucidate in more and more detail the metabolic pathways along which the subtle biochemistry of life proceeded and the research laboratories of industry could convert such knowledge into manufacturing processes. The intellectual climate presupposed that diligent research would lead – as, indeed, it did and continues to do – to useful innovation and economic advance.

What again changed the climate of thought was the elucidation by Watson and Crick of the structure of DNA – the chemical molecule of heredity – and the subsequent development of techniques by which pieces of the molecule responsible for particular characteristics could be transferred from one organism to another. This was genetic engineering. The new prospects offered seemed limitless so that many unfamiliar with the difficulties of scientific research jumped, in the intellectual

9

climate of heady excitement, to the conclusion that almost anything could be done immediately and hurried forward with hope and money. In reality, much is possible but it is important that the public should understand what the scientists can now do and the limits of what they may be able to do within a reasonable period of time. Thus this book attempts to put the present and future of biotechnology into perspective.

*Dr Magnus Pyke*

# Introduction

'Why trouble to make compounds yourself when a bug will do it for you?'
J. B. S. Haldane, 1929.

Over five decades ago Haldane, one of the most perceptive scientists of his time, encapsulated the logic behind what is now called biotechnology. Yet it is only in the last five years that the term biotechnology has escaped the confines of a few research laboratories. This book traces the scientific and technological discoveries that have led us to the brink of a new industrial revolution – the bioindustrial revolution. The myriad prospects of biotechnology are even more far-reaching than that of the silicon microchip. The microchip is essentially a device for handling information, while biotechnology can produce *materials* – from fuels to medicines, from foods to vaccines, and from chemicals to plastics.

The recent dramatic developments in biotechnology have stirred stockbrokers and company directors, beguiled politicians with the prospect of new sources of wealth, and inspired journalists to write of 'cures for cancer'. But the main story of biotechnology does not revolve around these people, nor even the scientists whose genius has brought them Nobel prizes. The real stars of biotechnology can only be seen with the help of a microscope – tiny microbes, and cells taken from plants and animals.

The innate skills of minute living cells are truly astounding. Millions of years of evolution have endowed them with a staggering versatility and resilience. Microbes can be found almost everywhere – in boiling water, locked in ice, immersed in oil. Some can feed on the apparently most unnutritious materials – petrol, wood, plastic, even solid rock. When, in addition, the substances microbes can manufacture are examined the immense potential of biotechnology starts to be revealed. Antibiotics, insecticides, fuels, dyes, industrial chemicals and vitamins are just a few of the multitude of valuable materials that can be obtained from microbes.

These facts would amply justify an intense interest in developing new industries employing workforces of millions of microbes busily manufacturing substances we need. However, the biotechnology boom has been truly sparked by the advent of genetic engineering. It is only a decade since scientists first discovered that they could graft totally foreign pieces of genetic information into microbes. The exploits of genetic

11

engineers have given a new twist to Haldane's words: if you can't find a bug that makes what you want, then create one that will!

In laboratories around the world genetic engineers have already designed microbes to manufacture dozens of potentially invaluable substances. Insulin produced by microbes has now been approved for use in diabetics in the UK and US. Interferon, which certainly fights viral diseases and may combat cancer, is now being tested on volunteers, while growth hormone, which reverses one of the major causes of dwarfism, will soon be available in large quantities from genetically engineered microbes. These are only part of the first wave of products from the fledgling genetic engineering industry. Other medical products set to follow include drugs to treat strokes, burns, nerve damage and, perhaps, even obesity. With all these medical possibilities on the horizon, as well as the production of fuels and chemicals for industry, it is not surprising that genetic engineering has received the lion's share of headlines in recent years. However, other dramatic advances in biology will prove equally influential in the bioindustrial revolution. Cell culture – the cultivation of fragments of plants or animals in the laboratory – has opened up countless exciting possibilities. The impact of research in plant cell culture may enable plant breeders to create new crops which grow more rapidly, require less fertilizer and thrive in poor soils. Monoclonal antibodies, the marker molecules manufactured by white blood cells, have already begun to revolutionize medical diagnosis.

As well as the allure of novel biological industries built on scientific discoveries of the last decade, the past and present of biotechnology have produced numerous triumphs. The flowering of the microbe-based antibiotics industry has already saved countless millions of lives. Indeed, humans have used microbes for millennia. A lunch of bread, cheese and beer is made possible only through the activities of microbes which turn milk into cheese, produce gases to make bread rise and convert the sugar in barley into alcohol. Until recently, the use of microbes was a largely hit-and-miss affair, for we lacked the knowledge to understand this microscopic world and the skills to manipulate it to our advantage. Now the efforts of thousands of scientists and technologists are being handsomely repaid, and enticing prospects for biotechnologists are to be found in almost every area of modern life.

### In medicine
New and improved treatments for the three major killers in developed countries: diseases of the heart and blood vessels, cancer and diabetes.

Better and cheaper antibiotics to counter the spread of infectious organisms that have developed a resistance to conventional antibiotics.

Vaccines to protect against viral diseases, such as hepatitis, influenza and rabies, and parasitic diseases, including malaria and sleeping sickness, which strike down millions of people each year.

Rapid tests that will aid doctors to make accurate diagnoses of many diseases.

Improved methods for matching organs for transplantation.

Techniques for correcting the body's chemistry to cure hereditary diseases, such as haemophilia.

**In agriculture and food production**
The creation of crops which make their own fertilizers, with immense savings in costs to the farmer.

Plants which can thrive on land that presently lies barren because the soil lacks water or is too salty.

Substances that can speed the growth of farm animals.

Vaccines to protect cattle from foot-and-mouth disease.

Cheaper forms of animal feed, consisting of microbes grown on waste materials.

**In energy production**
Renewable fuels, including methane and hydrogen gases, and fuel alcohol for domestic and industrial use.

Substances manufactured by microbes that will help to extract oil locked underground.

**In industry**
New sources of raw materials for the manufacture of plastics, paints, artificial fibres and adhesives.

Microbes which can extract metals from solid rock.

New systems for controlling pollution.

This list gives just a few of the benefits biotechnology can bring. It is not unreasonable for the public to treat the prophets of new technologies with a healthy scepticism and, indeed, the era of 'bio-hype' has arrived, with many extravagant claims being made for the future of biotechnology. Such inflation is both foolish and unnecessary; the realistic expectations for the bioindustrial revolution are quite impressive enough to command the attention of everyone who wants to know how our world will change in the next few years.

This book concentrates on the *possible* and the *actual*. Many of the biotechnological processes and products already mentioned have been *proved* to work. Some form the basis of vast industries which yield medicines, food and drink or prevent pollution of the environment. Others are presently in operation on a small scale in laboratories and within a short time will make a significant impact on our health, agriculture, energy supplies and industry.

13

Some of the milestones of biotechnology's past and some signposts to its future are listed in the table at the end of this chapter. The glittering prizes have naturally attracted financial and industrial analysts as well as a string of government-sponsored investigations. Predictions about the market for biotechnological products include: genetically engineered amino acids and proteins worth $5 billion by the year 2000; monoclonal antibodies worth $8 billion by 1992; pharmaceuticals worth $11 billion by 1990; chemicals worth $11 billion by 2000; agriculture and food processing worth $9 billion by 2000; energy worth $16 billion by 2000; and pollution control worth $48 million by 1991.* These amply illustrate why the corporate giants of today's industry are eager to take a stake in biotechnology. There is certainly more than the day-dreams of a few scientists involved when investment comes from companies such as Shell, Exxon, ICI, Glaxo, Grand Metropolitan, Standard Oil, Unilever, Allied Lyons, Cadbury Schweppes and Rank Hovis McDougall.

Biotechnology, like many new terms, means different things to different people. Some definitions are drawn so widely that they include conventional agriculture and animal breeding. At the other extreme, some popular accounts equate biotechnology with genetic engineering, ignoring numerous other equally exciting and rewarding techniques that now exist to turn our knowledge of biology to practical ends.

The crucial feature of biotechnology, as defined here, is that it makes use of *microbes*, or *cells obtained from plants and animals*, excluding activities which involve whole plants or animals, for example, growing wheat or raising cattle. Microbes are usually grown under carefully controlled conditions in large containers, sometimes huge metal vats with a capacity of 100,000 litres (22,000 gal.) or more. When supplied with the right nutrients, they can grow extremely rapidly. The total weight of microbes in a vessel may double in as little as twenty minutes. Thus, vast amounts of microbes can be obtained very quickly from a few starting organisms. Sometimes the biotechnologist's aim is simply to generate as many as possible at the cheapest cost. This biomass is widely used as a foodstuff for farm animals and, in some parts of the world, as a food for humans.

Most of the new biotechnological processes, however, are much more sophisticated. They aim to harvest certain valuable materials manufactured by the microbes. These include antibiotics, fuels and an enormous range of chemicals for industry. There are many materials of great value that, for all their versatility, no known microbe manufactures naturally, and this is where genetic engineering steps into the picture. The potential of genetic engineering is almost boundless and when its

* The sources of these financial estimates are respectively: Genex Corporation/US Office of Technology Assessment; International Resource Development; Business Communications Report; T. A. Sheets Company; International Planning Information Inc.; T. A. Sheets Company; Business Communications Co.

awe-inspiring power became apparent in the seventies, serious concern arose about the wisdom and safety of 'meddling with nature' in such a fundamental manner – altering the very genetic make-up of organisms. Alarming visions of mutant 'killer bugs' were conjured up. The scientists involved in this research were the first to pose questions about its possible dangers, and an intense public debate quickly followed throughout the world. Memories of the passions raised linger on in the minds of many, but it is not the purpose of this book to go into the complex details of these arguments. The final chapter briefly examines some of the new information that has been gathered over the last few years, which has convinced the vast majority of researchers – and the public bodies set up to oversee their work – that the dangers originally proposed do not, in fact, exist.

Today, as the bioindustrial revolution gathers momentum, we are faced with other questions of immense significance concerning the routes biotechnology will follow. Will there be further improvements in the health of people in developed nations with little attention to the immeasurably more severe afflictions of the developing world? Will increased fuel supplies only be created at the expense of diminished food resources? Will some nations be left behind in the rush towards biological industries and, if so, how will their economies be affected? How can research and development best be guided towards ensuring that biotechnology yields the most beneficial results?

As in the early days of the microchip revolution, the potential social and economic impact of new technologies can only be estimated by comprehending the underlying principles and their applications. Thus, the major part of this book presents specific examples of how biotechnology has already affected our lives and how its influence will become ever more pervasive in the next two decades. The aim is to provide readers with the information they need to assess the wider ramifications of biotechnology as they unfold.

The biotechnology boom of stock-market fever and screaming headlines has been fuelled to some degree by over-optimistic speculation. Fortunately, a healthy realism is now taking hold, and the days of bio-hype seem numbered. Biotechnology is emphatically not a fashionable catchword for interesting laboratory experiments, nor is it the key to instant megaprofits. Biotechnology will survive and thrive by translating knowledge drawn from many areas of science and technology into practical processes.

The problems today's biotechnologists face are sometimes complex and always challenging. The dramatic progress towards solutions to many of these problems – and the vast rewards for success – make biotechnology one of the most remarkable and fascinating endeavours of the eighties.

## Milestones in biotechnology

| | |
|---|---|
| Yeasts employed to make wine and beer | before 6000 BC |
| Leavened bread produced with the aid of yeasts | approx. 4000 BC |
| Aztecs harvest algae from lakes as a source of food | before AD 1521 |
| Copper mined with aid of microbes, Rio Tinto, Spain | before 1670 |
| Antoni van Leeuwenhoek first sees microbes with his newly designed microscope | 1680 |
| Louis Pasteur identifies extraneous microbes as a cause of failed beer fermentations | 1876 |
| Alcohol first used to fuel motors | approx. 1890 |
| Eduard Buchner discovers that enzymes extracted from yeast can convert sugar into alcohol | 1897 |
| Large-scale sewage purification systems employing microbes are established | approx. 1910 |
| Three important industrial chemicals (acetone, butanol and glycerol) obtained from bacteria | 1912–14 |
| Alexander Fleming discovers penicillin | 1928 |
| Large-scale production of penicillin begins | 1944 |
| Double helix structure of DNA revealed | 1953 |
| Introduction of many new antibiotics (streptomycin, cephalosporin, etc) | nineteen-fifties |
| Mining of uranium with the aid of microbes begins in Canada | 1962 |
| Brazilian government initiates major fuel programme to replace oil with alcohol | 1973 |
| First successful genetic engineering experiments | 1973 |
| Hybridomas which make monoclonal antibodies first created | 1975 |
| US National Institutes of Health introduces guidelines on genetic engineering | 1976 |
| Rank Hovis McDougall receive permission to market fungal food for human consumption in UK | 1980 |
| US Court decides that genetically engineered microbes can be patented | 1981 |
| Genetically engineered insulin approved for use in diabetics in US and UK | 1982 |
| First approval for release of genetically engineered microbes into the environment | 1982 |

## Signposts for the future

Genetically engineered growth hormone approved for
   treatment of dwarfism — 1984

Interferon used to treat some viral diseases
Monoclonal antibodies widely employed in diagnoses
Genetically engineered hepatitis vaccine introduced
New antibiotics produced by cell fusion
Commercial production of dyes and industrial chemicals
   from algae
Genetically engineered proteins used to treat heart attacks
   and strokes
Monoclonal antibodies employed to boost the body's
   defences against cancer and other diseases
New vaccines against foot-and-mouth disease
Growth hormones used to increase yields of meat and milk
   from cattle — mid-eighties

Raw materials for plastics industry obtained from microbes
Interferon employed to treat certain types of cancer
More industrial chemicals produced by microbes — late-eighties

Genetically engineered microbes help extract oil from the
   ground
Microbes widely employed to extract metals from factory
   wastes
Small-scale production of hydrogen from bacteria
Monoclonal antibodies used to guide anti-cancer drugs to
   cancerous tissues
New crops created by genetic engineering are able to
   manufacture their own fertilizers and resist drought and
   diseases — nineties

# CHAPTER ONE

# The chemistry of life
# and the key to biotechnology

The term biotechnology encompasses many activities, which have in common the fact that they all harness the fundamental abilities of living organisms. To understand, then, how they can be put to work and what biotechnology can do, it is important to be familiar with the basic chemical principles of life, in particular with the structure and function of proteins and of the genetic material, DNA.

One of the most impressive achievements of science has been the explosion of knowledge about the chemical composition of organisms and the way these chemicals interact to create the phenomenon we recognize as life. Probably the most powerful impetus given to biological research was the final acceptance last century that it is futile to search for 'vital forces' which distinguish living organisms from inanimate matter. The special nature of living beings does not reside in unique chemical principles, but rather in the immensely sophisticated way in which they utilize the ordinary laws of chemistry.

Organisms are sometimes compared to chemical factories. The strength of this analogy lies in its emphasis on the chemical nature of life – the fact that growth, development and reproduction all depend on chemical reactions. However, the analogy does obscure some of the most fundamental characteristics of organisms, many of which are directly relevant to biotechnology.

Perhaps the most notable feature of the living organisms is the sheer diversity of chemical processes they undertake. Most chemical factories are designed to convert specific raw materials into just a few products. Evolution has endowed organisms with the ability to take in a wide variety of raw materials (nutrients) and transform them into literally thousands of different types of materials, each with a particular biological role. Biotechnology draws its strength from these powerful chemical 'skills', developed over billions of years of evolution.

# The architecture of life

The world contains an astounding diversity of life forms, yet when we delve beneath the surface we discover that the millions of species that populate our planet have many features in common. The basic unit of biological organization is the cell. Plants and animals are all constructed from cells, each of which is surrounded by one or more thin membranes. Cells are very small – it would take 5000 human red blood cells to cover the dot on the letter i.

The plants and animals with which we are all familiar are composed of astronomical numbers of cells – the human body, for instance, contains about 100 billion cells. In these multicellular organisms, there are hundreds of different types of cells, each with its specific tasks: cells in the eye sense light, muscle cells provide the power for movement, and so on. This specialization of cells is a key feature of advanced organisms. Each type of cell contributes to the well-being of the whole organism and each depends on the others for its survival.

In contrast, most of the organisms which concern biotechnologists – the microbes – consist of only one cell. Each cell is an independent entity which can perform all the functions needed to keep it alive and allow it to reproduce. Thus, for these creatures the term 'cell' and 'organism' are synonymous (see Figure 1 and Plate 1).

The world of microbes includes several distinct types of organism. Three groups are especially important in biotechnology: the bacteria, the algae and the fungi. Bacteria generally have one of three shapes, rod-like, spherical or spiral, and most are between one and ten millionths of a metre in length. Algal cells tend to be slightly larger and one of their characteristic features is that, like plants, they can obtain energy from sunlight. The cells of fungi are often organized into large groups, as in mushrooms, but each cell is still capable of surviving on its own. Yeasts and bread moulds are both fungi.

Even though a bacterium may weigh only a millionth of a gram, its chemical versatility is quite dazzling. The cell consists of thousands of different types of chemicals, most of which are very complex. All of these substances must be constructed from the relatively simple building materials which the microbe finds in its environment. Complex chemicals cannot be built up from these simple materials in one step. Instead, each cell takes in raw materials and makes a succession of small modifications to them until the final product is completed. The network of chemical reactions by which cells convert an enormous range of substances into the materials they need for life is termed their metabolism. All of these individual chemical reactions must be harmoniously coordinated, and proteins called enzymes play a central role at every stage.

Protein molecules consist mostly of carbon, oxygen, hydrogen and nitrogen. All proteins are constructed from twenty different sorts of

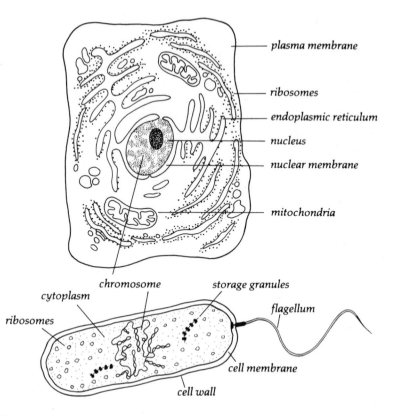

**Figure 1**
Despite the obvious structural differences between bacterial and animal cells their basic chemistry is very similar.

*Above* A typical bacterial cell. The cell is enclosed in a thin cell membrane, outside which is a more sturdy cell wall. These help the cell retain its shape and prevent the content of the inside (the cytoplasm) from leaking away. Many bacteria have a flagellum, which is used to move the cell around in a liquid environment. Storage granules contain food, and the ribosomes are globular structures on which proteins are manufactured. Finally, the chromosome contains the cell's genes, encoded in a DNA molecule which is tightly folded, like a bundled piece of rope.

*Top* A typical animal cell. The cells of plants and animals are much more complex and varied in appearance than those of bacteria. The chromosome is enclosed in a nucleus with its own membrane. The ribosomes are attached to a structure called the endoplasmic reticulum. Sausage-shaped bodies – the mitochondria – provide the cell with energy. The cell is enclosed by a plasma membrane which is the equivalent of the thin cell membrane in bacteria; an animal cell has no cell wall.

simple building block called amino acids. The staggering versatility of proteins arises from the fact that so many different shapes can be created by arranging amino acids in various ways. A chain containing only a dozen different amino acids, selected from the pool of twenty different sorts, can be strung together in billions of different permutations. Some proteins contain just a few dozen amino acid subunits, while others have over 200; most enzymes are composed of more than a hundred amino acid subunits. To make a protein, the individual amino acids are linked together in a chain, and the number and order of subunits in a particular type of protein give it its individuality.

Once the cell has assembled a chain of amino acids, the chain begins to fold back on itself. Only rarely do proteins exist as a straight line of amino acids; far more commonly, the chain twists and turns to create a complex three-dimensional structure (see Figure 2). This folding is not random, but is dictated by chemical forces, which depend on the sequence of amino acids in a particular protein. Thus, the final shape of a protein molecule is determined by the order in which its amino acids are strung together. Some enzymes and other proteins consist of more than one chain, and the chains for these are usually manufactured separately and later assembled to form the complete structure.

The importance of proteins can hardly be overestimated. The insight of the Dutch agricultural chemist Gerardus Mulder, who in the eighteen-twenties coined the word protein from the Greek for 'primary substance', has been amply vindicated by subsequent scientific investigations. The human body contains over 30,000 distinct types of protein. Each has a very specific use – for example, some give tendons their strength and resilience, some carry oxygen about the body, and others protect against infection. It is certain that many proteins still await discovery and, even so, a list of the known proteins would fill many pages of this book. At present, biotechnology is concerned with just a small fraction of these proteins, in particular those that act as enzymes.

## Enzymes – the biological accelerators

Enzymes are biological catalysts, a catalyst being any substance that speeds up the rate of a chemical reaction. Some chemical reactions take place rapidly and spontaneously, without a catalyst, as soon as chemicals are mixed. For example, hydrogen and fluorine molecules will combine very rapidly to form hydrogen fluoride. However, many reactions, including most that are important for life, take place only very slowly in the absence of an appropriate catalyst.

Although enzymes are grouped together on the basis of their ability to speed up chemical reactions, it is crucial to note that the different types of enzymes vary greatly in their structure and function. Each type of

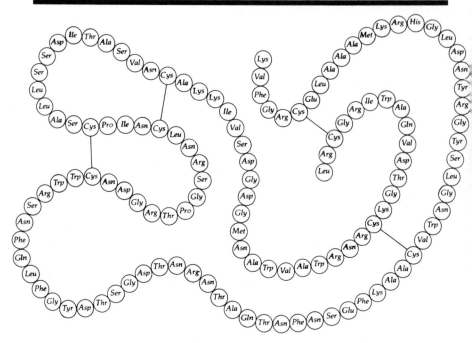

**Figure 2**
The enzyme lysozyme is constructed from 129 amino acid building blocks, represented here by circles. Each circle is labelled with a three-letter code according to the type of amino acid it represents, and the order of these is the same in every lysozyme molecule. Obviously, the three-dimensional shape of this molecule cannot be shown on a flat page; the molecule is actually folded into a form resembling a rugby ball. Lysozyme is found in human tears and is able to break down the cell walls of some bacteria. Alexander Fleming discovered this enzyme in 1922, seven years before he discovered penicillin.

enzyme has a particular molecular 'architecture' and most are able to trigger off only one particular type of chemical reaction. A typical cell may contain 1000 different types of enzymes and, between them, the many cells in a large and complex organism may have tens of thousands of distinct enzymes. So many are needed because the intricate task of keeping an organism alive and ensuring that it produces healthy offspring involves a vast array of chemical reactions, nearly all of which require the services of an enzyme. There are, of course, many identical copies of each type of enzyme in a cell.

Most enzymes are named after the type of chemical reaction they catalyse, and normally the suffix '-ase' is added to the name. Thus, the enzyme alcohol dehydrogenase is so named because it catalyses the removal of hydrogen from a molecule of alcohol. This enzyme is partly responsible for the fact that one eventually sobers up after drinking alcohol. However, by subtracting hydrogen from alcohol it produces acetaldehyde, a compound that induces a hang-over!

How enzymes act as catalysts is one of the most basic problems of biochemistry. Although many of the subtleties of enzymes remain hidden, much is known about the general features that give them their marvellous powers. The key is their three-dimensional shape. Every chemical compound has a characteristic shape, and an enzyme will interact only with those chemicals whose shape it 'recognizes'. The chemicals that take part in a reaction catalysed by a particular enzyme are termed the substrates of that enzyme.

The whole process by which an enzyme picks out its substrate or substrates works rather like a lock and key. On the surface of the enzyme there are clefts and crevices which match the bumps and protrusions on the substrate. Thus, when an enzyme and its substrate encounter each other they fit together. Any other compound (that is, one that is not a substrate for that enzyme) cannot act as a key to the enzyme lock – the key is the wrong shape. Once enzyme and substrate are locked together, chemical forces come into play which break and make various chemical bonds within the substrate, and alterations of chemical bonds are the essence of reactions.

Figure 3 shows, in outline, how one enzyme can recognize its two substrates and aid their transformation into two product molecules with different shapes and chemical properties. All of life depends on millions of such events taking place in an immensely complex, but highly coordinated system. Once the enzyme has performed its appointed task the products are released and the enzyme stands ready to repeat the whole sequence again, being in exactly the same chemical and physical state as it was at the beginning. This is characteristic of all catalysts and gives an enzyme molecule the ability to convert as many as a million substrate molecules every minute.

At first sight the extreme fastidiousness of enzymes – their rejection of all molecules except their substrates – may appear to be a disadvantage. In fact, it is crucial to their great power. Enzymes are not jacks-of-all-trades, but they are certainly masters of one. They possess both the speed of action essential for the cell and for the biotechnologist, and another, equally important attribute – precision.

An enzyme does not only select its substrates from the many materials that mill around it, but also ensures that the correct products are made. Most compounds in the cell can be altered in a variety of ways – adding an atom or two here, taking a couple off there, splitting the molecule down the middle, and so on. To form complex compounds, the cell must arrange for a series of small modifications to be made, one after the other,

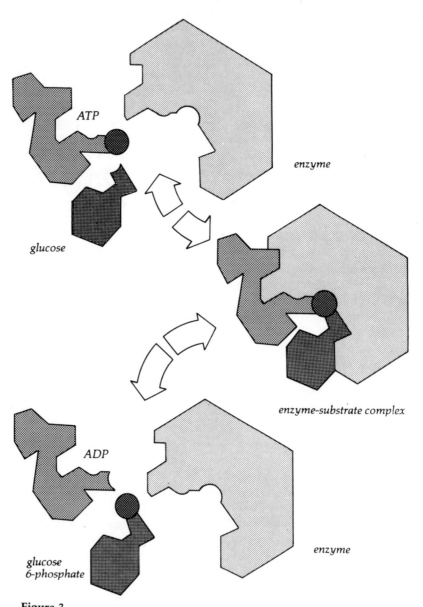

**Figure 3**
The lock and key model of enzymes. The two molecules (ATP and glucose) which are to react with each other fit into pockets in the surface of an enzyme. When they are brought close together by the enzyme, part of one molecule (the dark circle) is transferred to the other molecule. The products of the reaction (ADP and glucose 6-phosphate) are then released.

in a consistent and predictable manner. The mere accelerating effect of enzymes is not sufficient; each enzyme must deliver the correct, partly formed compound to the next enzyme in the series, so that it can perform its task. A motor-car assembly line provides a useful analogy. Simple components are modified and assembled in a strictly defined order to produce a complicated product. The speed at which each machine (enzyme) in the process works is obviously important, but it is equally crucial that a particular machine makes exactly the same product every time. If the products emerging from a machine vary, then they cannot be handled by the next machine in line and the process grinds to a halt. In the case of the cell, such a variation would lead to chaos (see Figure 4).

This ability of enzymes to channel reactions down particular pathways gives a high yield of the desired product and ensures that little of the precious starting materials is converted into unwanted, or even harmful, byproducts. In this way a highly complex network of reactions can go on constantly inside cells with predictable results, and predictability – at least at this molecular level – is the essence of life.

The finely tuned metabolism of cells can present problems for biotechnologists, however. Many biotechnological processes are designed to produce large quantities of a particular substance that cells normally manufacture in only modest amounts. An example is the

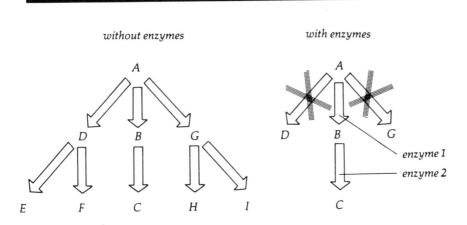

**Figure 4**
*Left* How the compound might undergo chemical reactions in the absence of 'directing' enzymes. Many different products result from the several different reactions that may occur.
*Right* Two enzymes 'channel' the reaction along a single path, ensuring that only one product is made – the product the cell needs.

production of lysine, an amino acid that is widely used as an additive to animal foodstuffs.

The vast majority of bacteria of the species *Corynebacterium glutamicum* produce only enough lysine to meet their needs. Built into the microbe's metabolic network is a system of feedback regulation by which the organism senses how much lysine it has available. If there is too little, a series of enzymic reactions takes place to replenish the supply; if there is enough lysine then the cell does not waste precious energy and raw materials in making more. Some of these bacteria, however, have regulatory systems which do not operate correctly and they manufacture much more lysine than they require (see Figure 5). Over-production of particular substances and the role of feedback regulation is of central importance in biotechnology.

# DNA – the spiral of life

In nature, like begets like. Poppies do not grow from rose seeds, nor do sheep give birth to calves. Obvious though these facts are, it is only in the last few decades that we have begun to unravel the underlying causes of this continuity in nature.

The fundamental unit of biological inheritance is the gene. Experiments over many years, particularly in the early years of this century, led scientists to develop some complex ideas about how genes operate and how they are passed from generation to generation. By studying creatures that reproduce rapidly, especially fruit flies, changes in their physical characteristics could be charted over many generations. This research eventually led to the notion that each gene is in some way responsible for making a particular type of enzyme. This 'one-gene, one-enzyme' theory was soon superseded by the more general theory that each gene is involved in the construction of one protein, which may be an enzyme or some other sort of protein. More recently, the theory has been further refined to state that each gene is responsible for making one type of amino acid chain. Some chains simply fold themselves to form an active enzyme; other enzymes consist of two or more chains.

Strange though it may seem, much of the pioneering work in genetics – the science of biological inheritance – was carried out by scientists who had only the fuzziest idea of what a gene actually *is*. Much was revealed about how these almost abstract entities affected physical characteristics, how they might be altered and how they are juggled during sexual reproduction. However, no-one knew precisely what sort of molecules genes are. This situation began to change in the late nineteen-forties when it became clear that genes are made of deoxyribonucleic acid, or DNA for short. DNA molecules are very long and sinuous (Plate 2) and it is now known that genes are simply parts of a DNA molecule.

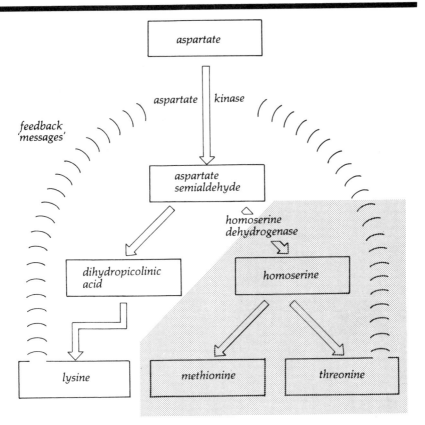

**Figure 5**

*Corynebacterium glutamicum* manufactures lysine from aspartate, a compound which also serves as a starting material for the manufacture of another amino acid, threonine. The rate at which aspartate is converted into these amino acids, via other chemical compounds, is regulated by the combined amounts of lysine and threonine in the cell. When there are sufficient amounts of both amino acids they act to inhibit (slow down) the enzyme aspartate kinase, which is responsible for channelling aspartate down the reaction sequences that produce lysine and threonine. However, in some forms of this bacterium the enzyme homoserine dehydrogenase is missing. This enzyme plays an essential role in the manufacture of threonine, but does not contribute to the production of lysine. Since the *combined* effect of lysine and threonine shuts down aspartate kinase, the first enzyme in the series, these defective bacteria continue to make lysine in large quantities since they are 'fooled' into acting as if the cell needs more lysine. As these bacteria need a certain amount of threonine to survive, biotechnologists supply the organism with a small amount of this amino acid – enough to keep it alive, but not enough to cause it to stop making lysine.

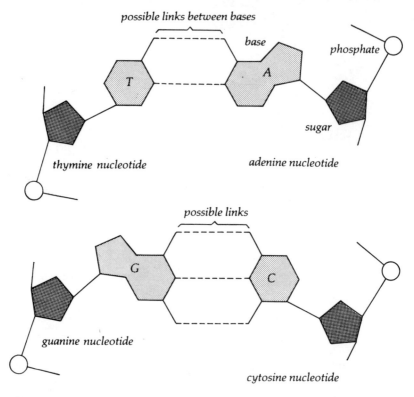

**Figure 6**
The four nucleotides found in DNA molecules. Each nucleotide is composed of three parts. The sugar and phosphate portions are identical in all of the DNA nucleotides. The base portions differ from one nucleotide to another. The figure also shows the potential links that each base can form – three each for G and C, and two each for A and T.

Genes are responsible for making proteins, and genes are made of DNA, but it would be slightly misleading to imply that DNA 'makes' proteins, rather it contains the *instructions* for manufacturing proteins. The laborious assembling of amino acids into a chain is the task of other components of the cell, with DNA directing the process. The unravelling of the mysteries of DNA and protein synthesis is one of the major scientific achievements of this century, perhaps even of any century, and has immeasurably widened the horizons of biotechnology. Once the relationship between DNA genes and proteins (especially enzymes) was elucidated it became possible to think about ways of inducing cells to make novel proteins by inserting the right pieces of DNA.

When cells are stained with special dyes and viewed with a microscope certain features stand out, among which are the chromosomes (Plate 3). The most important component of a chromosome is a single, huge molecule of DNA along which many genes are ranged. The number of chromosomes in a cell depends on the species of organism from which the cell is taken. Bacteria have only one chromosome, while human cells have forty-six. The chromosomes of bacteria float freely inside the cell, but in higher organisms, including plants and animals, the chromosomes are packaged inside a roughly spherical structure, the cell nucleus. The general principles concerning the structure of DNA and the manufacture of proteins are the same for all types of organisms. However, there are certain differences between those organisms that enclose their DNA in a nucleus (called the eukaryotes) and those that do not (the prokaryotes). These differences do not arise *because* higher organisms have a cell nucleus, but the presence of a nucleus distinguishes higher from lower organisms. At a molecular level, tiny yeast cells, which do have a nucleus, are more closely related to humans than they are to superficially similar bacteria.

Despite its immense size compared with other molecules, DNA has a remarkably simple structure. It consists of just four types of sub-unit called nucleotides, which are linked together in very specific orders to form chains. It is this ordering of nucleotides along the length of DNA molecules that serves as a code to convey all the information needed to instruct the cell to manufacture everything required for life.

Each nucleotide is composed of three sections. Two of these, a sugar called deoxyribose and a phosphate, are identical in each case; the third section, called the base, gives the four types of nucleotides their individuality. The shape of the nucleotides and their characteristic bases – adenine, thymine, cytosine and guanine – are shown in Figure 6, while Figure 7 shows how they can join together to form a chain which can be about 3 billion necleotides long in a human chromosome.

In the early fifties, only a few years after it was proved that DNA carries genetic messages, the structure of DNA was discovered and the clue to genetic inheritance was revealed. DNA takes the form of the now-famous double helix (see Figure 8), in which two chains of nucleotides twist around each other in a helix or spiral. The subunits in each chain are linked together as shown in Figure 7, but there are also links *between* the two chains. The bases of each chain face each other across the centre of the helix and they pair up in a very specific way. Adenine (A) and thymine (T) can each form two bonds, while cytosine (C) and guanine (G) can form three bonds each. Because they match up with each other in this way, the bases A and T are called complementary bases, as are C and G. This exact and unvarying relationship between the sequences of the bases on the two strands is used by the cell for two processes: to make copies of its DNA, and to assemble its proteins in an exact and reproducible way.

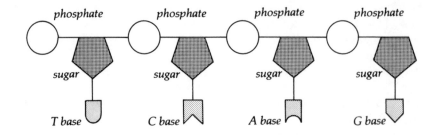

**Figure 7**
A chain of nucleotides is formed by linking each one through the sugar and phosphate parts of the molecules. The bases of the nucleotides branch off this backbone chain.

Even the most complex multicellular organisms grow from a single cell – a fertilized egg. This cell starts out with a complete set of chromosomes, half of which are provided by its female parent and the other half by the male. When it has grown to a certain size, this first cell splits into two, both of which also grow and split into two, until, eventually, the millions of cells that make up a mature adult have been formed. If, at each division, one cell took half the chromosomes (and hence half of the genetic information in DNA) after a few stages very little DNA would be left in each cell, the material having been distributed throughout many cells. Clearly, this would rapidly destroy the complex organization of life. Instead, the DNA in a cell duplicates itself just before the cell divides and, thus, the complete sets of chromosomes are available for the two new cells. This is where the double helix comes in.

Figure 9 illustrates how a molecule of DNA copies itself. The process can be viewed in terms familiar to photographers. Compare one strand of the double helix to a film negative and the other strand to a positive print. Clearly, both carry the same information or picture, but in different forms. If the positive and the negative are separated, the former can be used to make a new negative, while the original negative can be used to make a new positive. We now have two identical positive-negative pairs where only one existed before. DNA replication works much like this.

Perhaps the most elegant feature of life at the molecular level is the way in which the information for making proteins is encoded in DNA and read as instructions by other parts of the cell which assemble the proteins. A clue to the process is to be found in the emphasis placed earlier in this chapter on the specific *order* of the bases that make up DNA and the amino acids that make up proteins. When proteins are synthesized, the

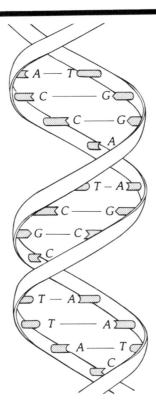

**Figure 8**

The DNA double helix. Two strands of nucleotides twist round each other and the strands are linked by bonds between bases in each strand. Guanine (G) always pairs with cytosine (C), while adenine (A) always pairs with thymine (T). The continuous 'ribbon' (the backbone) of each strand is composed of alternating sugar and phosphate sections.

order in which their amino acids are assembled is dictated by the order of bases in the relevant section of a DNA molecule – the gene for that protein.

The scientists who study these processes use the terms *transcription* and *translation* to describe them. These are among the most felicitous terms in biology because they make explicit an analogy with a written language of letters, words and sentences. The instructions in DNA are first transcribed, that is, written out in a similar form, and then translated, or converted into the language of proteins. Figures 10 and 11 show how the series of DNA bases in a small section of a gene are used to instruct the cell to assemble amino acids in a precise order to construct part of a protein molecule.

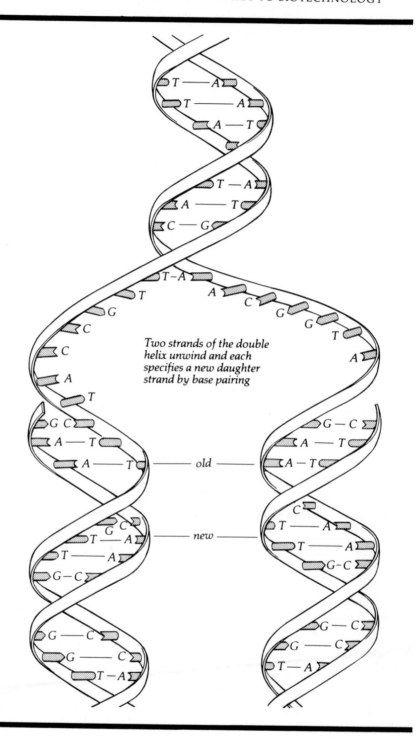

Two strands of the double
helix unwind and each
specifies a new daughter
strand by base pairing

old

new

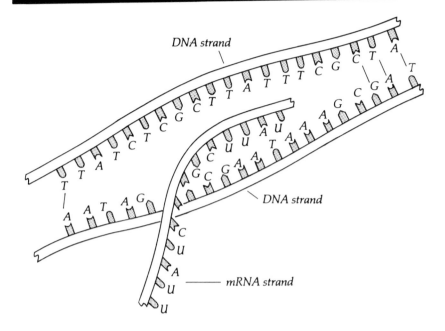

**Figure 10**
Forming a messenger RNA (mRNA) molecule. The backbones of both
DNA and RNA are represented by a single thick line, since only the bases
concern us now. The strands of DNA separate and the mRNA molecule is
built up according to the instructions contained in the sequence of bases on
one DNA strand. As in DNA, the C and G bases pair together, but in
mRNA a different base, uracil (U), replaces T as the partner of A. When the
mRNA molecule is complete, it peels off the DNA template and moves to
another part of the cell where its message is decoded to tell the cell how to
make the protein encoded in the DNA gene.

**Figure 9**
How DNA copies itself. The two strands of the double helix begin to
separate, breaking the bonds between the bases on each strand. The 'free'
bases now pick up new nucleotides from a pool of such molecules provided
by the cell. A pairs with T, and G pairs with C. Meanwhile, enzymes are at
work joining the newly selected bases into a chain. This process occurs with
both of the 'parent' strands and, thus, two identical DNA molecules are
created from one original. The specific pairing of the bases guarantees
faithful reproduction.

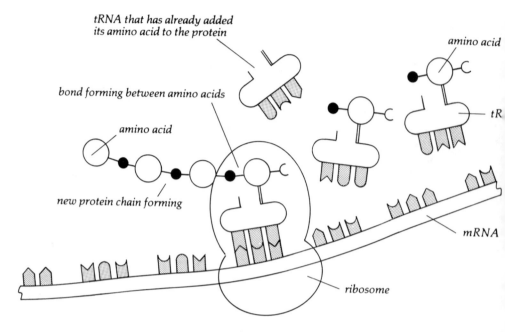

*tRNA that has already added its amino acid to the protein*

*amino acid*

*bond forming between amino acids*

*amino acid*

*tR*

*new protein chain forming*

*mRNA*

*ribosome*

**Figure 11**
The translation of mRNA involves dozens of types of molecules, among which are two other forms of RNA. Neither of these carries any genetic message, but they help the cell 'read' the information in mRNA. Once mRNA has been made, it leaves the nucleus (if that cell has a nucleus) and enters the main body of the cell. There, comparatively large globular structures, known as ribosomes, latch on to the wandering mRNA molecules. Ribosomes are built up from several sorts of protein plus a form of RNA called ribosomal RNA. Once the ribosomes have grasped the mRNA, the third type of RNA – transfer RNA (tRNA) – comes into action. There are many sorts of tRNA and each is able to recognize certain codons on the mRNA. Furthermore, each type of tRNA carries with it just one specific type of amino acid. The translation of the genetic code depends on the fact that one end of a tRNA molecule recognizes specific codons, while the other end of the same tRNA molecule carries a particular amino acid.

The mRNA is moved along the ribosome, much like a film moves across a ratchet in a camera. At each stage one 'frame' is made visible to the decoding centre, the 'frame' in this case being a codon. Many tRNA molecules mill around the ribosome and its mRNA. Each checks to see if the codon 'exposed' in the ribosome is one that it can match up with. If it is, the tRNA deposits the amino acid it has brought on to the last amino acid in the growing protein chain. The mRNA then moves another notch along the ribosome, exposing the next codon, and so the process continues.

First the message is transcribed into another type of molecule known as messenger RNA (mRNA for short). Chemically, mRNA is very similar to DNA; it too can recognize bases in a DNA strand and through this recognition process an mRNA 'working copy' of a gene is constructed. The process of translating the instruction on mRNA into the form of a protein now begins. The sequence of bases in an mRNA molecule specifies the sequence of amino acids in the protein. The words of the mRNA language consist of groups of three adjacent bases, together called a codon. Each codon tells the cellular machinery which synthesizes proteins that a particular amino acid must be incorporated into a protein at a specific place. Thus, step by step, proteins are constructed according to the instructions encoded in the DNA of genes.

The ability of organisms to live and reproduce depends absolutely on the manufacture of the correct proteins at the correct time. Today's biotechnology is being built on a deep understanding of the processes by which organisms, particularly microbes, achieve these dazzling feats of chemistry – and how we can capitalize on their skills.

# CHAPTER TWO

# Genetic engineering
# – reweaving the threads of life

'Genetically synthesized human insulin seems to be safe and effective in man.' These undramatic words brought news of a landmark in science and an event that could revolutionize medicine during the next decade.

In July 1980, seventeen volunteers received injections of insulin at Guy's Hospital, London. This seems unremarkable, for every day millions of diabetics are treated with insulin extracted from the pancreases of cattle or pigs, to control the amount of sugar in the bloodstream and fend off the debilitating effects of diabetes – a disease which is now the third largest killer in the developed world. However, the small group at Guy's were unique – they were the first humans ever to receive a substance that had been made by genetic engineering. Barely two years later, in September 1982, insulin from bacteria became the first genetically engineered material to be licensed for use in humans.

The importance of insulin derived from genetically engineered microbes lies less in the value of insulin itself (though this is immense) than in the fact that it helps vindicate the claim that microbes can be persuaded to make a wide variety of substances with medical and commercial value. It has proved that microbes can manufacture foreign (for example, human) proteins, and that these proteins are safe to use. Insulin is only the first of an avalanche of valuable substances which will be manufactured with the aid of genetically engineered microbes. The list grows monthly, and it now also includes interferon, growth hormone and blood proteins.

In the mid-nineteen-seventies an intense and fiery debate engulfed genetic engineering (also known as gene cloning, genetic manipulation, gene splicing or recombinant DNA research). With these entirely novel techniques, scientists can manipulate the very core of life – the DNA of which genes are composed. It is now possible to swap genes from one organism to another, inducing cells to manufacture materials they have never made before. Much of the storm over the wisdom of pursuing this awe-inspiring line of research revolved around the thorny issue of weighing possible risks (chiefly hypothetical killer 'superbugs') against the possible benefits, particularly in biological research and medicine.

In recent years this debate has cooled considerably and Chapter 9 presents some of the evidence and precautions which have led most biologists to conclude that the early fears about recombinant DNA research were overstated. Since genetic engineering is a key tool in biotechnology it is important to understand its basic principles and how these can be applied.

## Why turn to genetic engineering?

The possibility of transferring genes from one organism to another is an alluring prospect since genetic engineering could reduce the cost and increase the supply of an enormous range of materials now used in medicine, agriculture and industry. Furthermore, there are many substances that occur naturally in only small quantities which might well prove invaluable if they were available in sufficient quantities for their potential to be examined.

A major attraction of employing microbes as the factories for making these materials is that scientists and technologists have a great deal of experience in growing such organisms cheaply and efficiently on a large scale. Brewers and bakers have been doing this for millennia and the modern pharmaceutical industry has developed a new level of sophistication which will underpin much of the new biotechnological industries.

In essence, one of the major problems genetic engineering sets out to solve is that many types of cells cannot be grown outside their normal environment (at least not without an immense amount of difficulty), but very efficient techniques for growing microbes rapidly and cheaply have already been developed. Cells cannot be cultured from, for example, the human pancreas – the natural source of insulin – but genetic engineering can create microbes which can manufacture this material.

The basic principles of genetic engineering were developed only a decade ago, and since then there has been astounding progress which has provided us with a set of tools of quite remarkable power and sophistication. In outline, genetic engineering involves inserting new genetic information into an organism – usually a bacterium – to endow it with novel capabilities. It does not follow a single fixed set of procedures. The choice of method depends on which gene is to be transferred and what type of organism is to receive the new genetic information. The choice even depends to some extent on the personal preferences of the scientists involved.

The biotechnological applications of genetic engineering consist of four main stages: obtaining the gene which codes for the product the microbial factory is to manufacture; inserting the gene into the microbes; inducing the microbes to start synthesizing the foreign product; and collecting that product.

# First catch your gene

As with more familiar kinds of engineering, genetic engineering takes materials provided by nature, uses specialized knowledge and purpose-built tools to modify them in particular ways, and assembles the pieces to yield the final structure. However, genetic engineers are faced with particular problems arising from the fact that usually they cannot *see* the materials they are manipulating. Today's immensely powerful microscopes can provide a wealth of detail about the shape and structure of microbes, but give little information about the parts of the cell that are of central concern to genetic engineers – DNA and protein molecules. A human cell contains thousands of genes, each of which is the blueprint for manufacturing a protein, but even with the finest microscopes it is not possible to distinguish one gene from another. Thus, many of the subtlest tricks of genetic engineers are designed to reveal in other ways just what is going on inside the world of microbes, and what effects their manipulations are producing.

The first step towards creating in the laboratory a genetically engineered bacterium which will manufacture a human protein is to identify and isolate the gene from a human cell which codes for that protein. With a few exceptions – notably the cells involved in sexual reproduction – all cells of an organism contain the same genetic information, but they do not all use that information in the same way. The cells of different organs have different functions to perform, and this specialization among cells is possible because of a phenomenon known as gene expression. A gene is said to be expressed if it is employed to direct the construction of an mRNA molecule, which can then be used to manufacture the relevant protein. Genes can be turned on or off rather like a light bulb, particularly one fitted with a dimmer switch. Expression is not an all-or-nothing matter; some genes will be working at full tilt, some rather lethargically, while others are idle; it all depends on what proteins the cell needs at that time.

This difference in expression between genes in specific cell types is used by genetic engineers in what is, perhaps, the most elegant method of isolating the required gene. Rather than searching through the incredibly complex mass of genes encoded in DNA, this technique focuses on the cell's mRNA molecules. Thus, if a gene to make insulin is required, human pancreas cells are examined to find the mRNA molecules which have been copied from that gene. There will, of course, be other types of mRNA in that cell as well, but nothing like the multitude of diverse pieces of genetic information encountered in the cell's DNA.

The mRNA molecules must first be extracted from the cells. When cells are broken open, a great assortment of materials is released, including enzymes, other proteins, DNA, pieces of cell membrane, and mRNA. All these materials are separated from each other by treating the mixture with chemicals and subjecting it to a variety of physical forces. One of the most important steps is to put the mixture inside tubes which are spun round at very high speeds in a centrifuge. This separates different kinds of molecules according to their weights and shapes. Heavy globular materials quickly fall to the bottom of the tube while lighter, sinuous molecules are found higher up the tube.

The isolated population of mRNA molecules contains the genetic information needed to manufacture all the proteins that the cell was synthesizing at the time it was destroyed. This information must now be converted back into the form of the equivalent DNA sequences. (This DNA will, of course, only contain the far fewer distinct pieces of genetic information found on the genes the cell was using and not the thousands of other genes that were dormant or 'unexpressed'.)

Only a few years ago there was no known method for converting genetic information from an mRNA form to a DNA form. Much was known about the enzymes that help cells to make RNA copies of DNA genes, but information never seemed to flow the other way. Then came a revelation – some viruses can perform the trick. They do so with the aid of enzymes, reverse transcriptases, so called because they carry out the reverse of the operation that transcribes DNA into mRNA. Viruses, which are responsible for many diseases of microbes, plants and animals, are quite unlike any other entities. They are far smaller than even bacteria, and cannot be considered as truly living organisms since they are incapable of an independent existence. Viruses consist simply of a package of genetic material wrapped inside a protein coat. They reproduce by subverting the chemical machinery of host cells and forcing them to make replicas of the invading viruses. Certain types of virus employ RNA as their primary genetic material. Since all forms of life – from bacteria to humans – store their genetic information in DNA molecules, these RNA viruses have reverse transcriptase enzymes which convert their RNA genes into a DNA form, thus tricking the host cell into creating new viruses.

Mixing reverse transcriptase with the human mRNAs and the building blocks of DNA produces DNA copies of the RNA molecules. This DNA is constructed in the form of single strands, rather than the more familiar double-stranded helix. Other types of enzymes – DNA polymerases – are then used to convert single-stranded DNA into the double-stranded form by stringing together bases in a second strand according to the sequence of bases on the first strand. The double-stranded DNA so produced is called copy, or complementary, or cDNA, because it is a copy of the genetic instructions contained in the original mRNAs (an example is given in Figure 1).

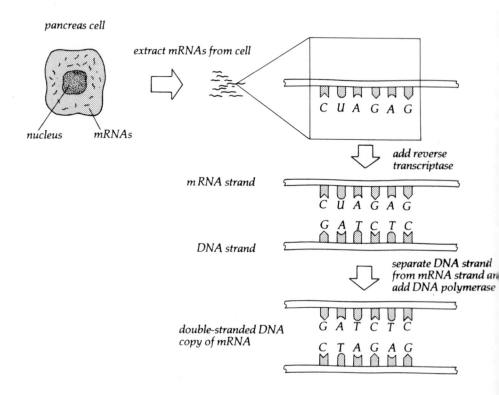

**Figure 1**

Human insulin is one of many proteins manufactured in the pancreas. Each protein has its own gene and its own type of mRNA. Pancreas cells are broken apart, releasing many sorts of mRNA, which are then separated from the rest of the cell material. Only part of the sequence of bases of a single mRNA is shown. The enzyme, reverse transcriptase, is added to the mRNA along with the nucleotide building blocks of DNA. This enzyme assembles the nucleotides in a string according to the order of bases on the mRNA – that is, it uses mRNA as a 'template'. The newly synthesized DNA is converted from single-stranded form into a double-stranded helix with the aid of another enzyme, DNA polymerase. This cDNA contains the same genetic information as the mRNA from which it has been formed. The cDNA is, in effect, a copy of a human gene.

Since the process started with many different types of mRNA, there are now cDNA copies of many different genes, among which are copies of the insulin gene.

# Plasmids – the 'magic' circles

All the genes essential for a bacterium's survival are carried on its single, large, circular chromosome. There are also much smaller circles of DNA inside some bacteria and these rings are known as plasmids. Plasmids are enigmatic structures and their functions are not entirely understood, although it is clear that many of them carry genes which enable bacteria to resist antibiotics. Plasmids have an odd relationship with the rest of the cell; most importantly for genetic engineering, they will often pass from one cell to another, even if the cells are of different species.

Therefore, if plasmids are taken from one set of bacteria and the human cDNA gene is 'stitched' into the plasmid ring, the plasmid's natural ability will allow it to enter bacteria and convey the human gene into its new home. A plasmid used in this way is called a vector, from the Latin for carrier or bearer. Certain types of viruses can also act as vectors.

To stitch the human gene into a plasmid we call upon the services of another type of enzyme, a restriction enzyme. All enzymes are remarkably precise tools, and in restriction enzymes their powerful ability to distinguish between similar structures is brought to a peak of perfection. Faced with a tangled mass of DNA, restriction enzymes scan the double helix until they recognize certain specific sequences of bases, and then make a precise cut across the two DNA strands. In the case of a circular plasmid molecule, this opens the ring ready for the insertion of the human gene.

About 300 different types of restriction enzyme have been found since the early nineteen-seventies. Each recognizes a specific sequence of bases in DNA and each makes its own characteristic type of incision. Figure 2 shows how three such enzymes operate. A crucial point is that they do not cut straight across the two strands, but instead create staggered ends. This leaves four bases hanging from each side of both cut strands. These bases are no longer paired with their normal companions and are, therefore, able to link up with any other piece of DNA which happens to have the same set of four hanging bases. Because two separate pieces of DNA cut with this restriction enzyme possess a mutual attraction, the enzymes are said to create sticky ends. These are the hooks upon which the human gene is hung (see Figure 3).

Meanwhile, in a separate set of test tubes and glass dishes, the human gene, in the form of cDNA, is prepared for its attachment to the opened plasmid. When the opened plasmid rings and the cDNAs are mixed some will join together forming new circles, each of which consists of the original plasmid with the human gene cDNA inserted into it. It is also possible for plasmids themselves to join together without capturing a human gene or, indeed, for two human gene cDNAs to link and form a circle. However, it is only the plasmid/human gene hybrid or recombinant molecules that are now of interest.

41

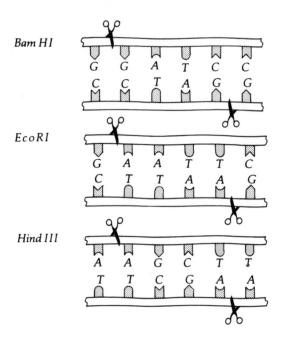

**Figure 2**

The restriction enzyme *Bam* HI (so called because it is obtained from the bacterium, *Bacillus amyloliquefaciens* H) recognizes the sequence of six base pairs shown and cuts across the two strands to create staggered ends with hanging bases. *Eco* RI (from *E. coli* RY13) and *Hind* III (from *Haemophilus influenza* Rd) operate in a similar way.

**Figure 3**

Creating recombinant DNA molecules. Plasmids are small circular pieces of DNA found inside some bacterial cells. The one on the left, named BR 322, contains one set of the sequence of six bases recognized and cut by the restriction enzyme *Bam* HI. This enzyme splits open the plasmid ring, leaving a set of four unpaired bases at each end of the molecule.

On the right is a cDNA molecule, to each end of which six linker bases are added. The same restriction enzyme is then used to cleave through the linkers, producing hanging (unpaired) bases which match those on the plasmid. The cDNA and the plasmids are now brought together and some will join, as shown, to produce a single circular molecule which incorporates the human gene cDNA. This is a recombinant DNA molecule. Because two pieces of DNA cut with this type of restriction enzyme possess a mutual attraction, the enzymes are said to create sticky ends.

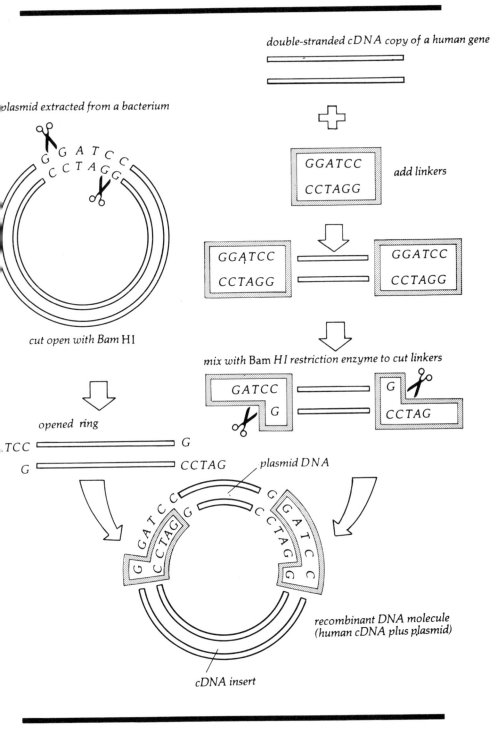

double-stranded cDNA copy of a human gene

plasmid extracted from a bacterium

add linkers

cut open with Bam HI

mix with Bam HI restriction enzyme to cut linkers

opened ring

plasmid DNA

recombinant DNA molecule
(human cDNA plus plasmid)

cDNA insert

At this stage the two pieces of DNA are held together rather loosely by only eight comparatively weak bonds between their sticky ends. A permanent link is forged with the aid of another enzyme, DNA ligase, which joins the backbones of each strand. Once these connections have been secured, the recombinant molecule is stable.

## A gene is cloned

Now, the recombinant molecule can be inserted into the bacteria that are to act as factories manufacturing the desired protein. By far the most popular choice of bacterial host is an organism known as *Escherichia coli* *(E. coli)*. The predominance of this bacterium in genetic engineering is partly an accident of history. Molecular biologists have studied this particular species for decades and, as a result, more is known about the inner workings of this microscopic creature than about any other organism, including humans.

The plasmids were chosen because they have an inherent ability to enter the cells of *E. coli*, and this invasion can be facilitated by adding a few simple chemicals to the mixture. Plasmids possess another property which is of great value to biotechnologists – they can make copies of themselves. Once inside a bacterial cell, a single plasmid may multiply itself to yield up to a few dozen identical replicas. If the plasmid contains a human gene, then that gene is copied along with the rest of the molecule. As the bacterium which harbours the plasmids is also growing and dividing – as often as once every twenty minutes – each daughter cell takes with it a few of the plasmids, which again reproduce themselves (see Figure 4). Before long a single bacterium will have given rise to millions of descendants. A population of cells all derived from a single ancestor is called a clone, and all cells in a clone have the same genetic make-up. Thus, within a day or so a single bacterium carrying a recombinant molecule will yield millions of identical cells, all of which contain the original human gene, and the gene is then said to have been cloned.

## The quest for the right bacteria

The genetic engineers now need to take stock of what types of bacteria are in the mixture of cells growing on a glass dish. The aim is to identify the very few bacteria which contain a recombinant plasmid with its human gene, for it is these bacteria which will be employed as microbial factories. They are confronted with three types of bacteria: first, there are bacteria that have been infected by a plasmid carrying a human gene (these are the ones the scientists are seeking); second, there are many bacteria which

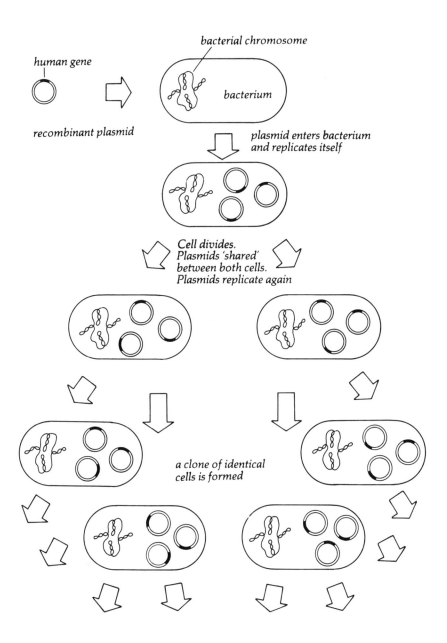

**Figure 4**
A recombinant plasmid containing a human gene enters a living bacterium
and is cloned.

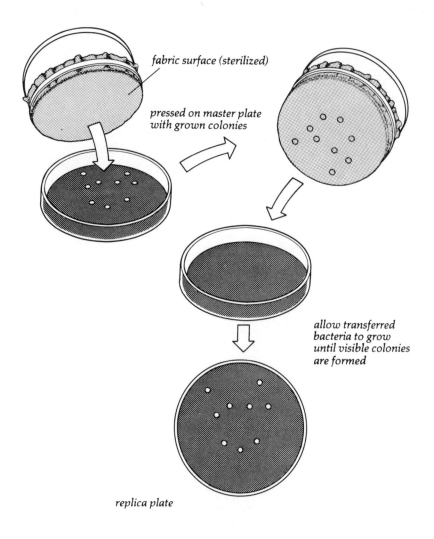

fabric surface (sterilized)

pressed on master plate
with grown colonies

allow transferred
bacteria to grow
until visible colonies
are formed

replica plate

**Figure 5**
Replica plating. Growing on a glass dish (Petri plate) of nutrients are many colonies of bacteria. All the cells in each clump are clones descended from a single parent cell. A pad of sterilized fabric is gently placed on the plate. Some cells from each clone adhere to the fabric, and this is then lifted off and transferred to another dish which is free of bacteria. The transferred bacteria soon multiply, producing visible colonies. In this way, two plates are obtained on which identical clones appear in exactly the same positions.

have picked up normal plasmids (that is, those which do not contain human genes); and third, there are bacteria that have resisted invasion by any type of plasmid. Some selection process is needed to identify the required bacteria and make visible the changes taking place in this microscopic world. A simple but invaluable technique which helps during the checking process is replica plating (see Figure 5). The mixed population of bacteria is smeared across a plate of nutrients to separate the individual cells. Each cell is allowed to multiply until it has produced a visible clone of cells. There will, of course, be many distinct clones on the plate. Each clone is then separated into two portions. One is retained to grow on its ideal nutrients, while various experiments are carried out on the other portion. Some of these experiments will kill some of the bacteria, but because their identical twins are kept alive and well in another place, copies of every clone are available for future investigations.

The bacteria which have no plasmids and those which have the normal plasmids need to be eliminated. Some types of plasmids carry genes which make bacteria resistant to certain antibiotics. One example, the plasmid coded BR322, is shown in Figure 6. An important feature of this

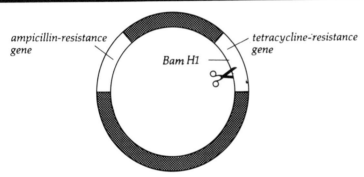

**Figure 6**
Plasmid BR322 is widely used in genetic engineering. It possesses two genes of particular interest. One (left) codes for a protein which enables any bacterium harbouring this plasmid to resist the lethal effects of the antibiotic, ampicillin. The other gene (right) similarly confers resistance against another antibiotic, tetracycline. This tetracycline-resistance gene happens to contain the series of six bases which are recognized and cut by the *Bam* HI restriction enzyme. If the plasmid is cleaved open with this enzyme and another piece of genetic material – for example, a human gene – is inserted, then the tetracycline-resistance gene no longer operates correctly. Any bacterium with such a recombinant plasmid/human gene molecule will be killed by tetracycline, but unaffected by ampicillin since the relevant gene for that antibiotic is undamaged.

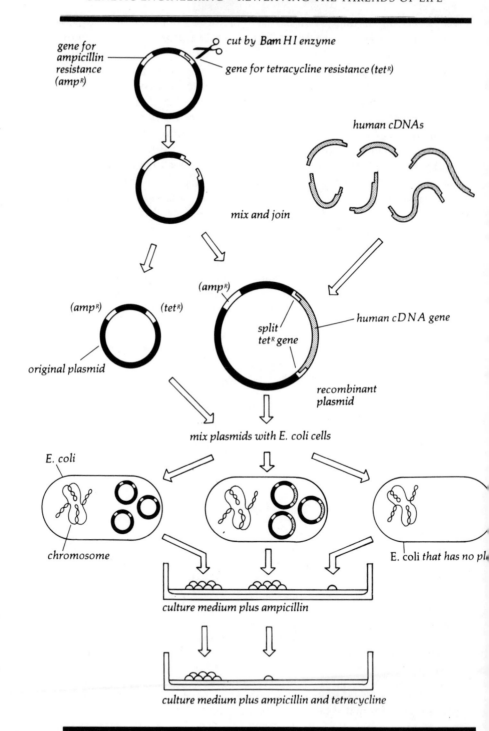

plasmid is that it contains genes which confer resistance to two distinct antibiotics, tetracycline and ampicillin. Any bacterium that contains the normal form of BR322 will be unharmed by these antibiotics. However, bacteria with recombinant plasmids will not be able to protect themselves against tetracycline. This forms the basis of a test (shown in Figure 7) to distinguish between bacteria with no plasmids (these will be killed by either antibiotic), those which have the normal plasmid (these can survive treatment with both tetracycline and ampicillin), and bacteria that have recombinant plasmids (these resist ampicillin but succumb to tetracycline).

The field has now been narrowed to the clones which contain a human gene of one sort or another. Since the whole procedure began with a mixed collection of mRNAs, only some clones will contain the required gene, while others will contain different human genes. There are a number of ways of isolating those with the required gene and a particularly elegant one is the radioactive antibody test. The success of this method hinges on two main facts: that molecules known as antibodies can recognize and latch on to specific types of proteins; and that tiny amounts of radiation can be detected.

Antibodies feature very prominently in biotechnology and this is just the first example of their many uses. Among the many thousand types of antibody molecules, there are a few that have a very strong affinity for a specific protein molecule. To search for bacterial clones that are producing insulin, genetic engineers rely on the help of antibodies which recognize insulin molecules and stick to them like labels. These labels,

---

**Figure 7**
Identifying the bacteria which harbour recombinant plasmid/human gene DNA molecules. Some of the bacteria that were mixed with plasmids will not have taken up a plasmid, and many of the remainder will contain unaltered plasmids which do not have a human gene. It is the third group of bacteria that are of interest – those that have been penetrated by recombinant plasmids. The clones of this last group of bacteria can be identified by a test that involves two sorts of antibiotic.

A portion of each clone is placed on a dish which contains nutrients and ampicillin. Clones that have no plasmids are killed. Those with any sort of plasmid survive as the ampicillin-resistance gene on the plasmid protects the cells. The surviving clones are then transferred to a dish which also contains tetracycline. This kills the cells which have recombinant plasmids since the introduction of a human gene in the middle of the tetracycline gene has destroyed its protective action. These are the clones required and their identical twins can be selected from the replica plate.

like most molecules, are far too small to be seen directly. However, 'tagging' these labels with radioactive atoms helps make them visible. These radioactively labelled antibodies emit very small amounts of radiation which produce a dark patch on photographic film, just as light darkens the film in an ordinary camera. Figure 8 shows how the combination of antibodies and radioactive labels pinpoints any colonies of bacteria that are manufacturing insulin.

## Turning genes on

The antibody test requires that the gene for insulin is present in a clone of bacteria, and that the bacteria are employing the information in that gene to manufacture insulin – that is, that the gene is expressed. For the biotechnologist it is absolutely essential that the foreign gene is expressed in the bacteria, since the whole point of the operation is to obtain the protein products of genes. To ensure that the gene is expressed, a few extra tricks are required.

The language of life is the same in bacteria and humans, but the dialects are different. While a set of three bases (a codon) instructs a bacterium and human cell alike to add exactly the same amino acid on to a growing protein chain, the so-called control regions which tell the two types of cell *when* (rather than how) to make a particular protein are different. Thus, to a bacterium, the natural control region of a human gene is totally meaningless, so it is simply ignored and no human protein is manufactured. The solution is to tag bacterial control regions on to human genes before they are inserted into their new home (see Figure 9). Then the bacteria see a familiar signal: 'start making mRNA from the gene next door', and that is precisely what they do, for they have no way of sensing that the adjacent gene is from a foreign organism.

The control regions in bacteria are of several different types. Some turn on or express the neighbouring gene only under special conditions. Others, which are associated with genes the bacteria need to express throughout their lives, continuously stimulate the cell to transcribe adjacent genes into mRNA. It is usually best to attach this latter type of control signal to the human gene so that as much of the human protein as possible is manufactured.

## Harvesting the products

In most cases the human protein remains inside the bacterial cells, so the only way to obtain it is to collect the cells and purify the required material from the great mass of debris that remains. The efficiency with which this can be done varies according to the particular microbes used and the protein produced. This is called down-stream processing and is a vital aspect of biotechnology. The general principles are the same for

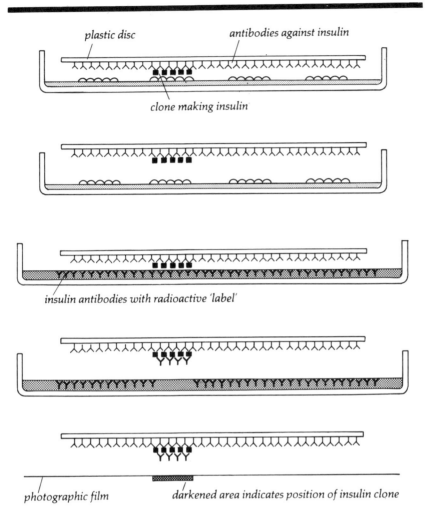

**plastic disc**  **antibodies against insulin**

**clone making insulin**

**insulin antibodies with radioactive 'label'**

**photographic film**  **darkened area indicates position of insulin clone**

**Figure 8**

Pinpointing the right clone. To search for the clones that manufacture, for example, insulin, antibodies which recognize insulin molecules are fixed to a plastic plate. When this plate is lowered on to a group of clones, it will pick up any insulin molecules that are present. Then the plate with its attached insulin molecules is placed on another layer of radioactively 'labelled' antibodies which can latch on to insulin molecules. These labelled antibodies are picked up *only* when they contact insulin. The whole plastic plate containing the antibody/insulin/radioactive antibody sandwiches is now placed on a photographic film. This becomes darkened wherever the radioactive antibodies are situated. By comparing the position of the dark marks on the film with the position of each clone on the original dish, the clones manufacturing insulin can be identified.

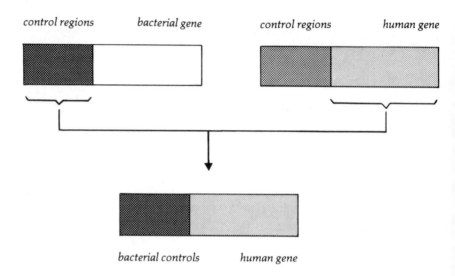

control regions      bacterial gene      control regions      human gene

bacterial controls      human gene

**Figure 9**
Typical bacterial and human genes both have control regions at one end in addition to the sequence of bases which specifies the order of amino acids in their proteins. Bacteria do not respond to human control signals. To persuade a bacterium to start making a human protein, the human control region must be replaced by one from a bacterial gene. Restriction enzymes are again invaluable in helping genetic engineers to snip DNA molecules at specific points. The fragments can then be rearranged to create a hybrid DNA molecule consisting of the gene which codes for the human protein plus the control region from a bacterial gene. Once inside a bacterium, these control regions instruct the cell to start making the human protein.

genetically engineered microbes as they are for natural microbes, and examples will be found in subsequent chapters. In essence, all the techniques make use of the particular protein's characteristic chemical and physical properties – its size, shape, electrical charge, solubility in water or other liquids, reactivity towards other chemicals, and so on.

There is a major service that genetic engineers can perform for the biotechnologists responsible for down-stream processing – bacteria can be tricked into secreting human proteins into their surroundings. Many microbes have efficient systems for transporting some of their natural proteins across their membranes and into the surroundings. This is necessary, for instance, in the case of enzymes that destroy antibiotics;

the bacterium cannot wait until the antibiotic has penetrated the cell before countering the threat. Proteins that are secreted from bacteria have 'for export' labels attached to them. These labels are specific sequences of amino acids and it is often possible for genetic engineers to ensure that the human protein is flagged in this way. The protein can then be isolated from the liquid surroundings of the bacteria rather than extracted from the bacteria themselves. The export label is later cut off the human protein.

# Jigsaw genes and junk DNA

In 1976 the summary of genetic engineering just presented would have been adequate. However, in the following year a remarkable discovery was made which bears directly on genetic engineering in the eighties – most human genes are in pieces. In bacteria and some types of algae (that is, the prokaryotic organisms) there is a direct relationship between the sequence of bases in a gene and the sequence of amino acids in its protein product. The first set of three bases indicates the first amino acid in the protein, the second set indicates the second amino acid and so on, until the last three bases signal the end of the protein chain. The genes of eukaryotic organisms (the group which includes yeasts, plants and animals) are often split. For example, the gene which codes for collagen – an important protein in tendons, bones and ligaments – is in more than fifty pieces. This does not mean that each part of the gene is on a separate DNA molecule, but that the DNA bases which code for the first stretch of amino acids are separated from the bases which code for the next stretch of amino acids by long strings of bases which do not contain information used in the construction of the collagen protein. These non-coding sequences are known variously as intervening sequences, introns or junk DNA. It seems that introns may help organisms to evolve new types of protein. By splitting the coding DNA into pieces it may be easier to juggle the basic units around to produce new proteins. This 'modular' theory is analogous to Lego building bricks; relatively few types of bricks (specific coding regions of DNA) can be rearranged to make many different structures (proteins).

Whatever the function of these introns is, they do not carry information which the cell normally uses to assemble amino acids into a protein chain. Eukaryotic organisms possess special enzymes which splice out the unwanted introns from mRNA which, when it is first transcribed from a gene, does contain introns. The resultant, shortened, messenger is called mature mRNA (see Figure 10) and it is this that is translated into protein. Similarly, it is mature mRNA that biotechnologists use to make cDNA copies of genes as discussed on p. 39. If they were to make a cDNA copy of the mRNA including introns and then insert that into bacteria (which

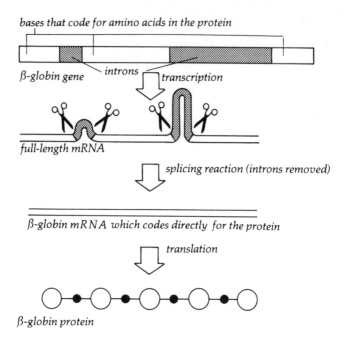

bases that code for amino acids in the protein

β-globin gene    introns    transcription

full-length mRNA

splicing reaction (introns removed)

β-globin mRNA which codes directly for the protein

translation

β-globin protein

**Figure 10**
When mRNA is first made from the gene of a higher organism it contains a copy of all the introns. Before the mRNA is employed to direct the synthesis of a protein, the cell removes the introns by a 'splicing' reaction. The shortened mRNA thus produced contains only the bases needed to direct the construction of a protein. It is this form of mRNA which is used to make cDNA, as described in Figure 1.

lack the ability to remove introns), the bacteria would translate the whole message into a string of amino acids which would be larger than the required protein and would not have the same properties.

Thus, a eukaryotic gene is like a bizarre manuscript. A few pages of exquisite prose are followed by meaningless letters and words before the next comprehensible passage. First, a messenger RNA copy of the entire text is made, then an expert editor excises the gibberish and reconnects the sections which convey useful information. The edited book is then ready to be 'read' on ribosomes and translated into a protein.

This fundamental difference between the organization of genes in animals and in bacteria severely limits the biotechnological applications of the aptly named shot-gun cloning methods. In the shot-gun approach, the entire mass of DNA from human cells is cut up into thousands of pieces by means of restriction enzymes. These pieces can be inserted into plasmids and cloned as the plasmids and their bacterial hosts reproduce, but no proper human protein is produced. At best, the end product is a protein with all the right amino acids in it, but interspersed with entirely irrelevant extra amino acids.

| first position | second position | | | | third position |
|---|---|---|---|---|---|
| | U | C | A | G | |
| U | Phe | Ser | Tyr | Cys | U |
| | Phe | Ser | Tyr | Cys | C |
| | Leu | Ser | Stop | Stop | A |
| | Leu | Ser | Stop | Trp | G |
| C | Leu | Pro | His | Arg | U |
| | Leu | Pro | His | Arg | C |
| | Leu | Pro | Gin | Arg | A |
| | Leu | Pro | Gin | Arg | G |
| A | Ile | Thr | Asn | Ser | U |
| | Ile | Thr | Asn | Ser | C |
| | Ile | Thr | Lys | Arg | A |
| | Met/ start | Thr | Lys | Arg | G |
| G | Val | Ala | Asp | Gly | U |
| | Val | Ala | Asp | Gly | C |
| | Val | Ala | Glu | Gly | A |
| | Val | Ala | Glu | Gly | G |

**Figure 11**
The genetic code. Each codon (sequence of three bases) in mRNA specifies a particular amino acid (indicated by its three-letter abbreviation). The genetic code shows this relationship, for example, the codon UCA specifies Ser, the amino acid serine. Three codons signal 'stop', the end of a protein chain. One codon, AUG, can indicate either the amino acid methionine (Met) or the start of a protein chain.

55

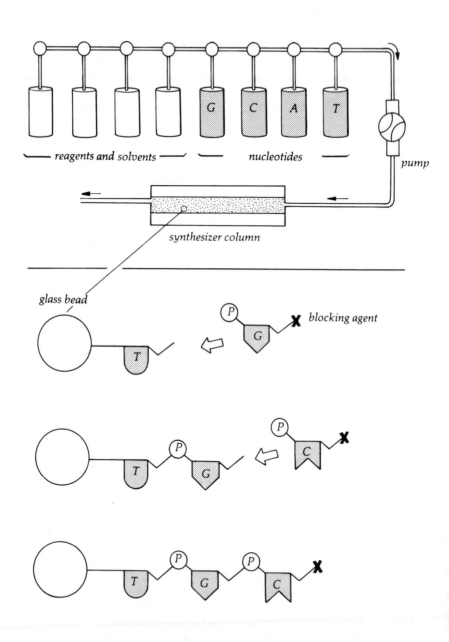

reagents and solvents — nucleotides —

pump

synthesizer column

glass bead

blocking agent

# Designer genes and gene machines

In 1953, Frederick Sanger and his colleagues in Cambridge crowned years of labour when they announced that they had worked out the sequence of amino acids in the insulin molecule. This was the first time that scientists had been able to decipher the order of building blocks in any protein. Today, a few thousand pounds will buy a machine which performs the complicated and repetitive chemical tricks required to analyse a protein. These amino acid analysers uncover the composition of a protein in a few days, while without them skilled chemists used to spend months or years getting the same results. Once the sequence of amino acids in a particular protein is known, it is an easy matter to look at the genetic code – charting which amino acids are specified by which three-base codons – and work out a DNA sequence that codes for the whole protein (Figure 11).

Another invention, the gene machine, can string together the building blocks of DNA in any order specified, and so allows the construction of genes from scratch, rather than teasing them out of cells (see Figure 12). With these machines, the era of do-it-yourself genes has arrived. Gene machines can link DNA subunits into a chain at the rate of about two per hour and speeds are increasing all the time. By 1982 it had become possible to synthesize a gene for interferon which consists of no less than 514 precisely ordered bases.

It is conceivable that, within a few years, a knowledge of how natural enzymes operate will enable biotechnologists to design totally novel proteins which perform specific chemical reactions with a very high efficiency. The scope of biotechnology in the chemical industry will be expanded immensely as molecular architects learn how to replace

---

**Figure 12**
A gene machine. Four separate reservoirs of the building blocks of DNA (the nucleotides A, T, C and G) are connected via tubes to a cylinder packed with glass beads. Suppose that the short series of nucleotides T–G–C is to be synthesized. The cylinder is first filled with beads with a single T attached, and then flooded with G bases from the reservoir. The right-hand side of each G shown in the picture is chemically modified or blocked so that it cannot join up with any of the other Gs in the surroundings. This prevents the sequence T–G–G . . . from building up. Any unbound Gs are now flushed from the cylinder and other chemicals added to remove the blocks. The cycle is then repeated with the introduction of C nucleotides into the cylinder and, thus, the sequence T–G–C is constructed on each glass bead. Once the desired sequence is finished, the chains of nucleotides are removed from the beads with the aid of chemicals. The whole process of pumping in the necessary chemicals at the right time is controlled by a microcomputer control system.

conventional catalysts employed in industry with proteins designed by humans. Gene machines will construct the genetic information required to induce microbes to manufacture these new catalysts.

Gene machines are already being put to practical use. The genetic instructions for the synthesis of the human insulin now on the market were, in fact, built up by chemists, rather than extracted from human cells, and so was part of the gene for human growth hormone.

Growth hormone is made in the pituitary, a small gland at the base of the brain, and secreted to reach all parts of the body where, as its name implies, it stimulates growth. Significant quantities of it are needed to treat some forms of dwarfism. Genetic engineers have managed to induce bacteria to yield this hormone, but to achieve this feat they had to introduce a few subtle refinements to the basic techniques. Since these extra tricks are likely to prove very useful in the manufacture of many different and valuable materials, it is worth looking at them in some detail.

The major part of the mRNA for human growth hormone codes for the 191 amino acids of the hormone itself. However, this hormone is secreted by pituitary cells, and a 'for export' signal is provided to ensure that this happens. This signal consists of an extra twenty-six amino acids at the beginning of the protein, and the instructions for manufacturing this signal are also encoded in the mRNA. The signal amino acids are excised from the protein when it leaves the pituitary cells, yielding the shorter active form of the hormone. This is both fine and necessary when the hormone is being made in human cells, but causes a problem when the gene is put into bacteria. The bacteria do not understand the signal's significance and do not remove the excess twenty-six amino acids. The result is a protein which is too large and would not perform its proper function when given to patients. Genetic engineers construct a partly artificial gene which will instruct bacteria to make growth hormone without the signal (see Figure 13). This gene can then be inserted into a plasmid, cloned and human growth hormone produced.

---

**Figure 13**
Assembling a human growth hormone gene. First, a cDNA copy of the mRNA for growth hormone plus signal is obtained. Ideally, genetic engineers would cut this in the place which corresponds exactly with the junction between the hormone and signal, but no known restriction enzyme can slice the DNA at that point. Instead, an *Hae*III restriction enzyme is employed. This cuts fifty bases off the cDNA, corresponding to the unwanted twenty-six amino acids *plus* twenty-four amino acids of the hormone itself. The missing genetic information is then replaced by synthesizing it in a gene machine and tagging it on to the truncated gene.

chemically synthesized piece of DNA
(72 bases = first 24 amino acids of growth hormone)

cDNA copy of growth hormone mRNA

use restriction enzyme Hae III
to remove signal bases and
those that code for the first
24 amino acids of the hormone

join.

complete gene without signal sequence

insert into plasmid

bacterial control region

recombinant
plasmid

growth hormone gene

clone in bacteria

complete growth hormone protein
(without signal sequence)

bacteria transcribe
and translate gene

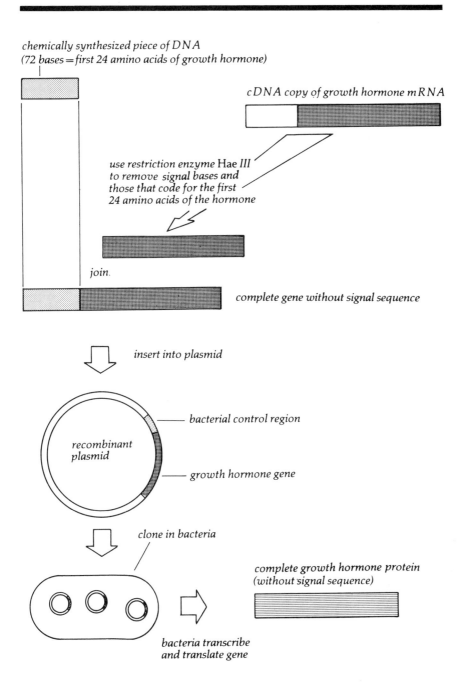

This kind of problem, and this type of solution, are frequently encountered in genetic engineering, particularly if a hormone is the target because all protein hormones are exported from the human cells that make them and, therefore, have some sort of signal. In the case of insulin, the situation is a little more complex still, but again gene machines provide a solution.

The insulin mRNA has an added level of sophistication. Not only does it possess a sequence which codes for the export signal at the beginning of the protein, it also codes for a stretch of thirty-five amino acids in the middle which are later cut out and discarded. Human pancreas cells can prepare insulin from the protein preproinsulin synthesized according to the instructions in the mRNA, but bacteria are totally baffled. They would meticulously translate all the mRNA into a single large protein which is of no medical use – it simply is not active insulin. Therefore, they must be given the genetic information in a way that does not require them to perform cut-and-join operations on the protein itself (see Figure 14).

This chapter has covered some of the immensely powerful and subtle tools of the genetic engineer's trade. The path is now clear for the insertion of almost any kind of gene into bacteria and other microbes, giving them the potential to manufacture a host of materials of great value in medicine, agriculture, the food and chemical industries, and in providing new sources of energy.

**Figure 14**
Genetic engineering of insulin. The mRNA of human insulin codes for four distinct groups of amino acids (top left): the A, B and C chains plus the signal sequence. When mRNA is translated a protein – preproinsulin – is formed which contains all four parts. To obtain active insulin, human pancreas cells first remove the signal sequence to yield proinsulin and then cut out the C chain, leaving only the linked A and B chains, insulin itself. Bacteria cannot perform this cutting and joining process. The production of genetically engineered insulin does not start with insulin mRNA, but with chemically synthesized pieces of DNA which code only for the A and B chains. The synthetic gene for the A chain is inserted into one batch of bacteria, and that for the B chain is placed in another batch. The A and B chains so produced are extracted from the E. coli cells, mixed and joined chemically to yield exact copies of human insulin molecules.

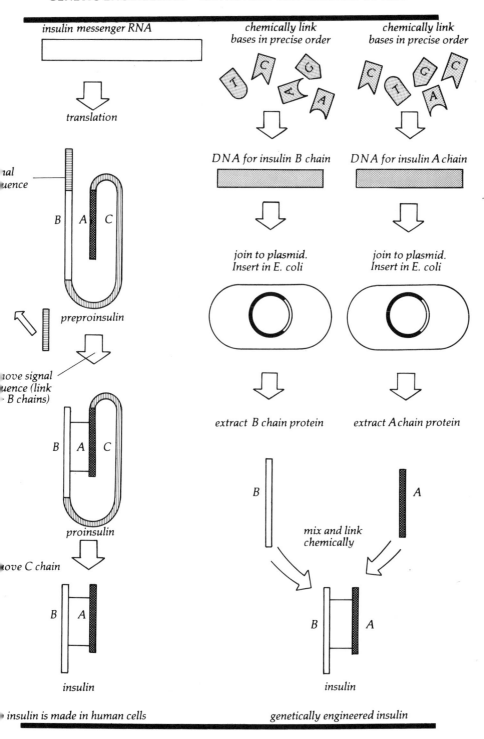

insulin messenger RNA

chemically link
bases in precise order

chemically link
bases in precise order

translation

DNA for insulin B chain

DNA for insulin A chain

ıal
ıence

B   A   C

join to plasmid.
Insert in E. coli

join to plasmid.
Insert in E. coli

preproinsulin

ıove signal
ıence (link
B chains)

extract B chain protein

extract A chain protein

B   A   C

B

A

proinsulin

mix and link
chemically

ıove C chain

B   A

B   A

insulin

insulin

insulin is made in human cells

genetically engineered insulin

# CHAPTER THREE

# Fermentation and selection
# – putting microbes to work

When studying the structure, metabolism and genetic information of microbes, it is rarely necessary for scientists to deal with large quantities of the organisms. A flask filled with about a litre of nutrients and growing microbes is usually ample. Biotechnology, however, sets out to produce large amounts of valuable materials – far more than even the most productive microbes can supply when grown in the confines of laboratory vessels. To create an efficient and economic biotechnological process it is very often necessary to scale up operations so that they work in huge metal vats. All the investigations of biologists who have discovered useful microbes, and the genetic engineers who have created new ones, would count for little in biotechnology if it were not for fermentation processes.

When words have both a scientific and a popular usage it is usually the former that is narrower and more specific. 'Fermentation' is an exception to this rule. In biotechnology, the term covers any process by which microbes are grown in large quantities to produce any type of material, not merely alcohol as in the common, restricted, sense of fermentation.

Success in biotechnology also depends crucially on discovering which type of microbe is best suited to large-scale fermentation processes and which is able to give the highest yields of the particular substance sought.

Even though the term biotechnology had not been coined at the time, the discovery and development of penicillin contains many of the most important elements which characterize biotechnology today. The story of how penicillin was discovered and how it came to be the first antibiotic manufactured in sufficient quantities to save the lives of millions, is, therefore, used to illustrate the general principles of selection and fermentation.

In 1928, Alexander Fleming was working at St Mary's Hospital, London, where he spent much of his time studying bacteria growing on small glass dishes in his rather untidy laboratory. One day in September he noticed that one of his dishes had been contaminated with a mould, *Penicillium notatum*. What really caught his attention was the fact that no bacteria had grown in the area around the mould (Plate 4). He surmised, correctly, that the mould had secreted some substance which inhibits the growth of bacteria, and he named the substance penicillin. Strangely, Fleming did not seize on this observation and follow it right through. He performed some more experiments to confirm his findings, and showed that the material was not toxic to mice. He did not, however, try to purify the active ingredient in the mixture of materials obtained from the mould, nor did he test the material's curative effects on infected mice.

In discovering that a certain type of mould makes some material which harms bacteria, Fleming supplied the first and crucial ingredient of most biotechnologies – he made an important scientific discovery about a property possessed by a living organism. Within a decade of 1928 the climate was ripe for his ideas to be picked up and put to practical use. Changes in scientific attitudes undoubtedly encouraged the development of penicillin as a therapeutic agent, but two other factors had a greater influence: with the outbreak of the Second World War there came a pressing need to find an effective antibiotic to treat the wounded; and a group of scientists and technologists with diverse skills intent on bringing the project to fruition were brought together.

Three main strands of research and development were intertwined in the penicillin story. First, the best type of mould had to be found; the mould used in large-scale production of penicillin had to manufacture as much of the antibiotic as possible and be fairly easy to handle in large quantities. Second, methods for separating penicillin from all the other substances produced by the mould were required. Third, it was necessary to design vessels in which the mould could be grown and from which the valuable harvest of penicillin could be recovered.

The first stages in the development of penicillin were carried out in Oxford by Howard Florey, Ernst Chain and their colleagues. They began work in 1938 and by 1940 American scientists had become heavily involved in the work. In total, several hundred scientists in thirty laboratories took part. This endeavour was pursued with such enthusiasm and skill that by 1944 there was enough penicillin to treat all the serious British and American casualties during the invasion of Europe.

Samples of the mould that happened to fall on Fleming's dish had been kept alive in St Mary's Hospital. Unfortunately, this particular mould does not produce very much penicillin, so the search was on to find other, slightly different moulds that would yield more. By 1951, it had become possible to obtain 60mg of penicillin from a 1-litre (0.22-gal) flask full of mould and liquid nutrients. Since then further improvements have been introduced which allow 20 g of penicillin to be recovered from the same volume of material – 10,000 times as much as could be produced in Florey's laboratory.

New and sophisticated techniques for separating penicillin from the mass of other materials in the flasks played a vital role in this remarkably improved yield, but equally important was the work of scientists who managed to improve the mould. The use of the word 'improve' in this context, and in biotechnology in general, is obviously anthropocentric; it means that some characteristic of an organism, in this case penicillin production, has been altered in a direction that suits our purposes. The change is unlikely to be of direct benefit to the organism concerned, for once the mould is able to manufacture enough penicillin to keep bacteria away from it, it gains nothing by escalating the submicroscopic arms race to a level at which it can kill the bacteria ten times over (Plate 5). Biotechnologists must create an artificial environment which encourages organisms to overproduce (from their own point of view) the materials we demand.

The techniques for improving an organism can be divided into two broad categories: genetic engineering, which, of course, was unknown until the seventies; and conventional techniques, which played a central role in the penicillin story and continue to do so in most areas of biotechnology today. Unlike genetic engineering, by which desirable characteristics can be introduced into organisms in a fairly controlled manner, conventional techniques for improving organisms must rely on more random procedures, making use of what nature provides. This does not mean that scientists are reduced to being mere observers of nature's activities, simply picking up the unconsidered trifles she casts out. Man, not nature, writes the rules of this evolutionary game.

The key to finding the organisms which are most suitable for a particular application lies in the existence of variations in nature. Not all individuals of the same species are identical. We can see this by observing the plants and animals around us. Variations also occur in microbes, and to a greater extent. The appearance of two cats may differ considerably, but their basic biochemistry is very similar. Both require the same substances in their food to survive, and both consist of very much the same chemical compounds in about the same amounts. The variations between two microbes of the same species are likely to be greater. If the differences are sufficiently interesting and well defined, the two microbes will be said to be different strains of the species.

The early workers on the penicillin project searched among many individual *Penicillium* moulds to see if a strain that manufactured an unusually large amount of penicillin could be found. This search was very successful, but required an enormous amount of painstaking and tedious work. Samples of mould were collected from as many sources as possible – one of the best was found growing on a cantaloupe in a New Jersey market. Samples of each mould were grown in the laboratory and tested to see how much antibiotic they produced. The labour involved in this screening process was immense but the rewards, when they came, were even greater.

Once the promising strain of *Penicillium* mould was unearthed, the scientists set about improving it further by inducing mutations in the organism. The differences between strains of *Penicillium*, including their ability to manufacture penicillin, arise from the fact that each has a slightly different set of genetic instructions. A mutation is simply a change in the normal genetic make-up of an organism. Mutations occur quite naturally – indeed, without them evolution would be impossible, for it is on slight variations between individuals in a species that natural selection acts. Some mutations will be immediately lethal for the organism involved; a few will be beneficial; and some will be more or less neutral, conferring no clear advantage or disadvantage on the organism possessing the mutation.

However, natural mutations occur infrequently. In the quest for a microbe that is more useful to humans, the rate at which mutations arise is increased, thereby increasing the variety of strains of the species available. It is of no concern to the biotechnologist that most of the individual moulds do not survive as long as one is created that fits the bill – in this case, by producing more penicillin.

By 1951, Fleming's original mould, *Penicillium notatum*, had been discarded as a potential large-scale source of penicillin and had been replaced by a similar species, *Penicillium chrysogenum*. Scientists began to search for the most suitable strain of this species, and to increase the natural rate of mutation they treated the mould with X-rays, ultraviolet light and the poison, nitrogen mustard. These harsh treatments induced mutations and, occasionally, threw up an improved strain. This strain would then be subjected to another series of experiments. In over twenty such steps, the strain in commercial use today was created.

# From laboratory to industry – the critical transition

In their quest for the best strains of *Penicillium*, scientists usually grew the mould in small glass or earthenware vessels. With such small-scale, labour-intensive operations there is no hope of producing enough penicillin to treat the vast number of patients who can benefit from the drug. To meet the massive demand for penicillin, the pharmaceutical companies of the eighties have installed gigantic metal vats with capacities of up to 100,000 litres (22,000 gals, see Plate 6).

To design the best fermentation processes, biotechnologists must pay careful attention to many factors, including the supply of the best nutrients, preventing contamination and controlling the fermentation conditions, such as temperature and acidity.

### Nutrients
If the *Penicillium* mould is to grow and provide penicillin, it must have access to a variety of nutrients which supply it with energy and the raw materials it needs to synthesize the compounds that make up its cells. In

terms of bulk, the major components of nutrients are sources of carbon and nitrogen. These provide both energy and building blocks for the cell's compounds and, since almost all of the cell's constituents contain carbon, the demand for carbon-containing nutrients is high. Two major factors influence the biotechnologist's choice of nutrients – economic and biological.

In laboratory experiments, scientists are at liberty to try out a vast range of potential nutrients to see which suits the microbes best. In an industrial operation, however, cost constraints are far more formidable. The prospects for any biotechnology are bleak if it becomes apparent that the organism requires expensive nutrients. Fortunately, this is not often the case and, indeed, some of the most successful biotechnologies utilize waste materials from industrial, agricultural or domestic sources.

Penicillin is still something of an enigma. There is much detailed information about the way it destroys bacteria by attacking their cell walls, but no-one is quite sure just *why* the *Penicillium* mould manufactures penicillin. The most obvious explanation is that it uses penicillin to keep foreign microbes at bay, preventing them from competing for the precious food resources. However, for most of its life-span *Penicillium* produces almost no penicillin at all; the synthesis of the antibiotic seems to be turned-on around the time the mould cells stop growing. Biotechnologists must overcome this and persuade the mould to start making penicillin as quickly as possible. Despite the lack of agreement about the biological reasons for *Penicillium*'s apparent reluctance to make penicillin, biotechnologists have been remarkably successful in achieving their goal.

*Penicillium* can use many sorts of compounds as sources of energy and carbon, most notably sugars of various kinds. One sugar in particular, glucose, is consumed very eagerly and the mould grows rapidly, but does not produce much penicillin. It was discovered that another sugar, lactose (which is found in milk), is less 'digestible' to *Penicillium* so the mould grows more slowly but produces much more penicillin. The pharmaceutical industry now feeds mixtures of glucose and lactose to the moulds, thus maintaining the optimum balance between cell growth and penicillin production.

Investigations into the most suitable nutrients are key factors in all biotechnological developments and sometimes, as in the case of penicillin, a stroke of luck can have a major impact. When corn (maize) is processed to manufacture starch, a liquid by-product known as corn-steep liquor remains. At one time this material was dried and sold as cattle feed, a means of disposing of the waste that was, at best, only barely profitable. For many years the US Department of Agriculture sought a more profitable use for corn-steep liquor, but to no avail. However, during the Second World War some of the Department's scientists were closely involved in the development of penicillin, and they decided to see if the *Penicillium* mould could grow on corn-steep liquor. The results were startling; not only did the mould happily consume this waste material, but yields of penicillin more than doubled. It is now known that corn-steep liquor contains an excellent mixture of carbon compounds, but more importantly, one of these compounds is a precursor of penicillin – a substance that the mould can use directly to build up penicillin molecules. While such serendipity cannot be guaranteed to bless other, newer biotechnologies it certainly serves to illustrate the pleasant surprises nature can spring upon aspiring biotechnologists.

Most of the multitude of compounds inside a cell are composed of a skeleton of carbon atoms with various other types of atoms attached. Clearly biotechnologists must ensure that these, too, are supplied to the growing organisms. In practice, this creates few difficulties. Hydrogen, the most abundant element in the cell, is present in almost every type of carbon compound and, of course, in water. The other elements of major importance in life can easily be provided in the form of simple, cheap compounds – nitrogen (used, for example, to construct proteins, DNA and RNA), phosphorus (DNA, RNA and cellular membranes), sulphur (many proteins), plus smaller quantities of iron, potassium, sodium, zinc and other metals. As the cellular chemistry of micro-organisms is so versatile, most can synthesize everything they need from very simple starting materials.

## Oxygen – the energy liberator

All the most familiar forms of life need oxygen to survive. Inside cells, oxygen is combined with molecules derived from foodstuffs, liberating energy which can be used to maintain all the cell's vital functions. Many of the microbes used in biotechnology, including *Penicillium*, also require oxygen. (Some do not – indeed some microbes are actually killed by oxygen, the most notorious example being *Clostridium botulinum*, the cause of botulinus food poisoning.) Controlling the amount of oxygen in a microbe's environment is hence often vital.

When *Penicillium* mould grows in small containers it can obtain all the oxygen it needs directly from the air; oxygen in the atmosphere diffuses into the mass of growing cells. However, oxygen cannot diffuse over large distances and cells near the bottom of the huge fermentation vats used in industry would die for lack of oxygen unless special arrangements were made to keep them well supplied. For this reason, air is pumped into the base of fermentation vessels.

It is relevant to note here another of the significant breakthroughs which enabled penicillin production to reach its present level of efficiency. *Penicillium notatum*, the first penicillin mould, only grows well when it is near the surface of a vessel. Clearly, there would have been little point in constructing vessels several metres deep if the mould inhabited only the top few centimetres. Fortunately, the strains of

*Penicillium chrysogenum* discovered in the forties not only make more penicillin, but also thrive when submerged.

## Maintaining an equable environment

A key factor that favours biological industries over many chemical methods of producing the same materials is that the former do not require high temperatures. Metallic catalysts often require temperatures of 100°C (212°F) and more to work efficiently. Most biotechnological processes operate best at 30–50°C (86–122°F), about the temperature of the human body. This brings many advantages, in particular it is rarely necessary to burn expensive fuel to keep fermentation vessels at the optimum temperature.

However, there is a drawback: it is usually essential that the temperature of the microbes is kept within a narrow range. The vessel's temperature must be monitored constantly, and a little heating or cooling applied as necessary. Living organisms are intolerant of extremes of temperature – say, below minus 10°C (14°F) and above 100°C (212°F) – because of the nature of their chemical compounds. If proteins, DNA or many of the other molecules essential for life are heated, their structure becomes disrupted and they no longer operate correctly. On the other hand, when the temperature drops, the catalytic effect of enzymes diminishes very rapidly. The almost incredible efficiency of enzymes has the penalty that they work well only at or very near a certain temperature. This temperature varies from species to species; polar fish are adapted to life at or below 0°C (32°F) and their enzymes can cope with the cold, whereas bacteria which inhabit hot geothermal springs have enzymes that remain active at temperatures as high as 85°C (185°F)*. The ability to maintain a constant body temperature is thought to have been one of the most important factors that allowed the mammals to supplant the reptiles as the dominant animals on land. In most mammals, the body temperature varies by only a degree or so, regardless of the external temperature. This means that their enzymes can work efficiently at all times, allowing them to remain active rather than dropping into a reptilian torpor at night or in cold weather.

*Penicillium* generates heat as it consumes nutrients. If the mould is grown in a small container, this excess heat is rapidly dissipated through the walls of the vessel. However, the large vessels used in industry must be cooled to prevent the heat building up and the mould being harmed. Figure 1 shows cooling jackets encircling the fermentation vessels and pipes and valves through which acid or alkali may be added to the fermentation medium.

---

* Recently bacteria have been discovered in volcanic vents at the bottom of the ocean. These organisms live under very great pressure and at temperatures of about 300°C. It is not yet known how they survive in such extreme conditions.

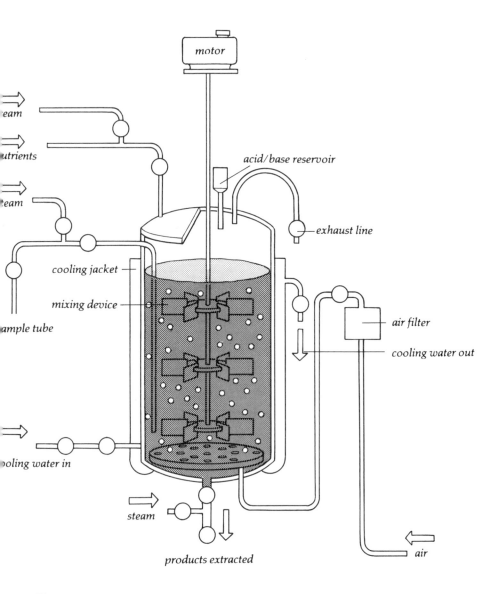

**Figure 1**
A typical fermentation vessel. The microbes are grown on nutrients placed
in the vessel at the start of the fermentation. The vessel is cooled by a water
jacket. Air is pumped into the bottom of the liquid, and acid or alkali added
as necessary. A stirrer keeps the contents well mixed. Steam lines are
provided so that the vessel can be sterilized after each fermentation batch.

Anyone who has made their own wine will know that it is vital to maintain the correct degree of acidity (pH) if a palatable drink is to result. Just as organisms have a preferred temperature, so a certain degree of acidity or alkalinity suits them best. In fact, penicillin production is particularly sensitive to the acidity of the environment – too much or too little causes the yield of antibiotic to fall considerably.

## Sterilization – keeping invaders at bay

From the cold, dark depths of the ocean to the steaming heat of geyser pools, microbes can be found everywhere on Earth. The resilience, diversity and versatility of the microbial world are crucial to the use of microbes in biotechnology, but their very hardiness does bring disadvantages. The vast majority of biotechnological processes use only a single species. There is little point in seeking out the best strain of *Penicillium chrysogenum* and then trying to grow it in a vessel that is teeming with other, quite different microbes. These intruders may compete for the available nutrients, slowing the growth of *Penicillium*. They may even excrete some dangerous toxin which will contaminate the product. Careful sterilization of all equipment and materials introduced into the fermentation vessel is, therefore, essential. This sounds easier than it is in practice. Potentially harmful microbes lurk everywhere and fastidious cleanliness is required to eliminate them.

Louis Pasteur was instrumental in delineating this central principle of biotechnology. In the eighteen-sixties, the production of alcohol from sugar beet was an important industry in the French town of Lille. Occasionally, for reasons quite incomprehensible to the manufacturers, the vats of sugar beet and yeast stubbornly refused to yield alcohol, becoming filled with all manner of foul-smelling substances. Pasteur discovered that the fermentation process went wrong when the vats became infected with other microbes. Ever since then the dangers of microbial contamination have been well understood, but infections still cause problems from time to time.

The microbes used in most biotechnologies are particularly at risk from being overwhelmed by competing organisms. This is because the strains employed have been artificially selected by humans on the basis of certain characteristics they possess – most obviously, in this case, greater production of penicillin. Any organism which overproduces (from its own point of view) any substance is very likely to be at a disadvantage among other organisms. This phenomenon is also encountered, for example, in farming. Wheat has been bred to exhibit characteristics which benefit humans, including high yields of seeds and short stems. These plants would fare badly among the natural competition in the wild, and it is the farmer's job to see that they get special treatment. In the same way, biotechnologists must ensure that their microbes are protected from competition.

This competition may come not only from foreign species (many of which are unaffected by penicillin), but also from related strains of *Penicillium*. A rogue mould may appear in the fermentation vessel, either by infection from outside or by a mutation in the production strain. If this rogue is more efficient in utilizing the available resources – perhaps by ceasing to overproduce penicillin – its descendants will soon outnumber those of the original strain. Once this has happened the vessel must be cleaned out, resterilized and the process started again with a fresh batch of the good strain of *Penicillium*. It is for this reason, among others, that most biotechnologies operate using batch processes. The whole system is set up and the microbes allowed to grow and produce the desired material (this takes about nine days for penicillin fermentation). The vessel is then emptied, the product purified (about fourteen hours for penicillin) and the vessel cleaned before a new batch is started.

Clearly, there are advantages to be gained from a shift towards continuous processes, which could operate for weeks or months with nutrients being gradually fed into the system and products being drawn off in a steady stream. Much effort is being put into the development of continuous systems in several areas of biotechnology.

**Purifying the product**

When the fermentation process is complete, the vessel is full of a thick broth of microbial cells, some unconsumed nutrients and dissolved penicillin. The extraction and purification of fermentation products varies widely according to the particular process. In the case of penicillin, a fairly small molecule which the cells excrete into the surrounding liquid, the problems are not too difficult to overcome. First, the cells are filtered from the penicillin-laden liquid and discarded. When certain potassium compounds are mixed into the liquid, the penicillin forms crystals which settle to the bottom of a container and can be collected. This basic form of penicillin, named penicillin G, may then be chemically modified to form a wide range of semi-synthetic penicillins with names such as ampicillin and methicillin. The chemical modification of antibiotics produced by microbes is a common feature in the pharmaceutical industry; biotechnologists and chemists cooperate to produce many different antibiotics, each with a particular application.

In other biotechnological processes it may be necessary to break open the microbial cells to release the desired product. Most enzymes, for example, remain locked inside the cell and are not excreted naturally. If it is necessary to disrupt the cells, the process of purification of the desired product is made more complicated by the presence of large amounts of cell debris. This may add substantially to the costs of the whole operation. It is likely, however, that before long genetic engineers will help reduce costs by inducing microbes to start secreting the required products into the fermentation liquids.

# CHAPTER FOUR

# Biotechnology and disease
# – prevention, diagnosis and cure

The introduction of penicillin and a host of newer antibiotics has lifted the scourge of many infectious diseases and saved millions of lives. Now, biotechnologists are paving the way for a massive assault on many more of the world's most devastating diseases, including cancer, diabetes, hepatitis, malaria and sleeping sickness. Less common, but equally dangerous inherited diseases, such as haemophilia, will also be tackled with the aid of biotechnology.

The medical applications of biological industries are exciting, and progress is astonishingly fast. Ten years ago many current techniques existed only in the realms of science fiction, and many more will become facts in only a few years' time. Many diverse aspects of biotechnology are being drafted into the fight against disease. In particular, naturally occurring and genetically engineered microbes can be employed to manufacture drugs, vaccines, hormones and enzymes; new tools, such as monoclonal antibodies, should aid diagnosis and therapy; while cell fusion could provide novel and powerful antibiotics.

This chapter examines the contributions biotechnologists can make in the prevention, diagnosis and cure of three groups of diseases: those that are caused by an invasion of the body by viruses, bacteria and other micro-organisms; those that result from some imbalance in the body's natural chemistry; and those whose causes are less well understood, including heart disease and cancer.

## Fending off invaders

In the eighteen-sixties and seventies, two of the greatest scientists of the time, Louis Pasteur in France and Robert Koch in Germany, were hard at work establishing a theory that has probably had more impact on modern medicine than any other. Between them, Koch and Pasteur proved that certain diseases of humans and other animals – including tuberculosis, anthrax and cholera – are caused by particular, identifiable microbes.

With the benefit of hindsight, this germ theory of disease may seem obvious, but at the time the notion was little short of revolutionary. Although Pasteur and Koch were not the first scientists to propose such a theory, their work provided the conclusive proof and dispelled the air of mystery that had previously shrouded ideas concerning the nature of disease. Before the germ theory took root, only the imagination of those who investigated diseases seemed to limit the possible 'causes' – miasmas in the air, supernatural influences, and character defects of the sick person had all been invoked to 'explain' diseases.

Once the germ theory had become entrenched in medical thinking, a new and more scientific era of diagnosis and treatment could begin. We now know that many common diseases are caused by bacteria, viruses and fungi, and that diseases spread as these are passed from person to person, either directly or indirectly. An immense amount of effort is now devoted to tracking down and identifying infectious microbes. Once the culprit has been unmasked, specific means of combatting it can be sought.

The study of microbes, microbiology, revealed that by no means all of them are harmful. This is fortunate since all of us play host to literally millions of microbes, and we would not survive long if most were not benign. With the advent of biotechnology, our relationship with microbes is entering a new phase – some microbes have been put to very positive medical uses, particularly in the production of antibiotics.

## Antibiotics – the proving ground of biotechnology

Today there are about 100 different antibiotics available for use on humans. Large though this number is, it is only a small fraction of the 5000 or so compounds isolated from microbes which have been shown to kill or disable other microbes. The large number of antibiotics that are not used in medicine are rejected for a range of reasons: some produce too many harmful side effects; some are too expensive to manufacture on a large scale; and some simply cannot do a specific job as well as another, readily available antibiotic.

The four major classes of antibiotics – the penicillins, the tetracyclines, the cephalosporins and erythromycin – are worth over US $4 billion in bulk sales each year, and all are superb examples of the art of biotechnology. Although the details of how each group came to benefit humans differ, the same general principles outlined for penicillin in Chapter 3 apply. The reason for the development of this range of antibiotics, and for the quest for new ones, can be seen in the story of the cephalosporins.

In 1945, Guiseppe Brotzu, a Sardinian professor of bacteriology, found a micro-organism in the sea near a sewage outfall. This fungus, a species of *Cephalosporium*, produced a substance that killed a wide range of bacteria. Brotzu did not have the facilities to analyse this substance, so he sent his organism to Oxford. There, in the laboratory of Howard Florey where the early penicillin work had been done, Guy Newton and Edward Abraham discovered that the fungus manufactured a novel type of penicillin, which they named penicillin N. Then in 1953 they made another discovery of much greater importance – this organism also manufactured another antibiotic, which they called cephalosporin C.

The immediate advantage offered by cephalosporin C was that it could kill bacteria that had become resistant to penicillin. The phenomenon of resistance to antibiotics has proved to be the major driving force behind the quest for new antibiotics. When penicillin was first introduced it appeared to be a wonder drug and, indeed, it stopped many of the commonest dangerous bacterial infections in their tracks. However, within a very few years doctors found to their dismay that some infections that had previously succumbed rapidly to a course of penicillin now managed to survive – the bacteria had become resistant to penicillin.

Resistance to a particular antibiotic can appear in a population of bacteria in a variety of ways, including the transfer of antibiotic-resistance plasmids from other species or strains. The mere fact that an antibiotic comes into common use perversely (from our point of view) encourages the appearance of resistant bacteria. For example, the early penicillin-resistant bacteria appeared to have 'started' making an enzyme called penicillinase, which attacks the antibiotic before it can do its job. In fact, there may always have been some individual cells making small amounts of the enzyme, but they were vastly outnumbered by the penicillin-sensitive type. As these latter were exterminated by penicillin, the resistant strains were able to thrive in their place.

Cephalosporin C was developed to tackle diseases, such as pneumonia, caused by penicillin-resistant Staphylococcus, but before cephalosporin was ready to be used on a large scale these bacteria had been largely conquered by one of the semi-synthetic penicillins, methicillin (produced by chemically modifying the basic penicillin molecule obtained directly from Penicillium moulds). Cephalosporins did, however, come into their own in the nineteen-sixties when they began to be employed to fight other types of bacterial infection, and they now form a crucial part of the armoury of anti-microbial compounds.

Both penicillin and cephalosporin are members of the group of antibiotics known as beta-lactams, after a characteristic type of chemical ring structure they possess. They operate by preventing certain bacteria from building proper cell walls.

Streptomycin, on the other hand, belongs to the group of antibiotics known as aminoglycosides. These work by preventing certain bacteria from manufacturing their proteins. Once inside the bacterium, streptomycin disrupts its ribosomes, the small globular structures on which the genetic information carried by mRNA is translated into proteins.

This antibiotic was discovered by Selman Waksman, one of the great figures in the history of microbiology, and his colleagues at Rutgers University, New Jersey. They spent years studying microbes from the soil, and in 1944 they announced that they had found a new antibiotic and named it streptomycin, because it came from the filamentous microbe *Streptomyces griseus*. Streptomycin proved particularly valuable because it attacks microbes which are unharmed by penicillin and cephalosporin. In particular, it revolutionized the treatment of tuberculosis: patients were no longer condemned to years of slow therapy in sanitoria, since the new drug could often effect a cure within months.

It is an odd fact, which has never been properly explained, that the majority of antibiotics, including streptomycin, are produced by a rather narrow range of organisms, collectively known as the actinomycetes. In appearance, actinomycetes are similar to the moulds that make penicillin and cephalosporin. The branching network of filaments in actinomycetes are composed, however, of bacterial or prokaryotic cells, whereas moulds consist of more complex, eukaryotic cells.

The alarming spread of resistance to several kinds of antibiotics demands that the quest for new forms continues. In the short term, no great advances can be expected through the type of genetic engineering already discussed – that is, the insertion of one or two genes which instruct a cell to make some protein it would not normally produce. None of the commonest antibiotics are proteins and they are not, therefore, the *direct* products of genes. Most antibiotics are constructed inside the cell by a chain of discrete chemical reactions, each of which is catalysed by a separate enzyme and, of course, each enzyme is made according to the plan encoded in its own gene. It is a daunting task to find out exactly how a particular antibiotic is created, what enzymes are involved and which genes underlie the whole process.

However, genetic engineering might be used to create modified antibiotics. There are many examples of related antibiotics used in medicine which are chemically very similar. A classic example is the penicillin group – all have the same basic structure, but the presence of a variety of comparatively minor chemical modifications results in a range of drugs with different uses. Today, only the basic structure of penicillin is obtained directly from the moulds; purely chemical methods are then used to tinker with the molecules to obtain the many types of penicillin. This can be done with some ease for the penicillins, but modifying some other types of antibiotic is more difficult. Possibly microbes could do the job instead.

The microbes would need to be persuaded to use certain enzymes they already possess to modify the standard antibiotics they normally produce. For instance, cells use enzymes called methyl transferases to swap a methyl unit (one carbon plus three hydrogen atoms) between pre-existing molecules. These methylation reactions are vital for many chemical processes within the cell. If antibiotic-producing cells could be induced to employ their methyltransferases to tack on methyl units to their antibiotic molecules, a new antibiotic would be produced with different and perhaps even more useful properties. Although such modified antibiotics are perhaps more likely to come through genetic engineering, the technique of cell fusion offers another possible avenue.

## Cell fusion – new microbes from old

Very few biotechnological processes employ 'natural' strains of microbes – that is, those found in the environment. Such 'wild-type' strains may be admirably adapted for survival in their normal habitats, but will probably yield relatively little of the substances we desire and may not be suited to the environment of a fermenter. Remember, for example, that natural strains of *Penicillium* mould produce only one ten-thousandth of the amount of penicillin made by the strains now employed in the pharmaceutical industry.

When seeking to produce a new strain of microbe, biotechnologists may rely on evolution by artificial rather than natural selection. For example, in the development of high-yielding strains of *Penicillium* mould, mutations were induced by treating the microbes with chemicals or radiation (see p. 65). This is a completely random process and there is no means of knowing what alterations will occur in the microbes' genetic material. Genetic engineering is much more specific, for it attempts to introduce a specific gene into the microbes. The technique of cell fusion lies at an intermediate position on this scale of random to specific genetic modification; its results are more predictable than random mutations but less so than genetic engineering.

The term cell fusion may be unfamiliar, but there is at least one very good reason why it is central to all our lives – each of us was created by a process of cell fusion. Sexual reproduction takes place when two cells carrying different genetic information join together, or fuse. The resulting combination of genes, some from the egg cell and some from the sperm cell, yields a new individual which is different from either parent and, indeed, is genetically unique. This constant reshuffling of genes from generation to generation produces much of the diversity among individuals of a species upon which natural selection operates.

Most microbes, however, predominantly reproduce themselves by simple division, in which a cell splits to give two identical offspring. Thus, in the absence of any random mutations, no new combinations of genes are obtained and the next generation will have exactly the same properties as the previous ones. Clearly this is a great advantage for biotechnologists once the required strain has been found, but it does not aid the search for new and better strains. The technique of cell fusion allows the generation of novel combinations of genes within microbes by joining two cells together.

The principles are quite straightforward. The outer, tough membrane surrounding bacterial cells is stripped away, usually with the aid of enzymes, to leave the cell contents packaged in the much more delicate inner membrane. These fragile forms of cells, called protoplasts, can then be induced to combine by adding certain viruses or chemicals (Plate 7).

Cell fusion creates hybrid or recombinant cells, which contain genetic material from two or more cells. Cells to be fused may be different strains of the same microbial species or even entirely distinct species. The great advantage is that novel mixes of genetic material can be obtained – combinations which would be found only rarely in nature or not at all. Apart from the creation of modified antibiotics mentioned earlier, cell fusion may also produce new antibiotics by activating 'silent' genes.

It is already known that the actinomycetes group of bacteria makes many hundreds of different antibiotics and more are being discovered each year. It is quite likely that these organisms have the inherent ability to make an even wider range of antibiotic substances, some of which might be medically useful. According to this theory, the genes which instruct the cell to make these, as yet undiscovered, antibiotics are 'silent' or unexpressed so that no mRNA is made according to their instructions and, thus, no proteins are synthesized. As mentioned on p. 38, the majority of any cell's genes are turned off at any one time. If the organism has no need for the protein encoded in a gene it does not normally waste its energy by making it. The problem is to persuade the actinomycetes' cells to turn on their silent genes so that their products can be assessed. It is reasonable to assume that the genes must be active under *some* conditions – temperature, nutrient supply or other external factors – for if the organism never used the genes they would probably have been cast aside during its evolution. One could invest a great deal of effort in trying to grow the cells under an enormous variety of conditions in the hope that some trigger for the activation of the genes will be found. However, cell fusion offers a short cut.

By no means all of a cell's DNA is used to carry codes which instruct the cell how to make certain proteins (see p. 53). Some stretches of DNA perform a control function, allowing the cell to turn on or off the protein-coding genes according to its changing needs. Many of the most exciting and awe-inspiring discoveries in basic biology over the last few years have been concerned with the control of gene expression. The stunning and subtle complexity of gene-control systems is now being rapidly revealed, although doubtless many more surprises await discovery. The essence, so far as this potential application of cell fusion is concerned, is that the control sequences of DNA do not 'know' what genes they are controlling. In normal cells, of course, the control systems have evolved in such a way that they turn their respective genes on and off according to the cell's needs for the protein encoded in that gene. If, however, a control sequence is moved from its natural position to a place where it controls an entirely different gene, it will switch its new companion on and off according to the control sequence's perception of the cell's need for the protein made by the gene with which it is *normally* associated.

Cell fusion is one way of bringing together unaccustomed partners – protein-coding genes and control regions. The aim is to attach one of the unexpressed antibiotic genes to a control region which will turn it on under the conditions found inside fermentation vessels. Under the invigorating influence of its new control region the previously silent gene should then become expressed – mRNA being copied from the gene and enzymes being synthesized which the microbe may use to manufacture new and useful antibiotics.

When cells are fused, particularly if they are closely related strains of microbes, pieces of DNA from each of the original cells will sometimes swap places. In particular, a control region from cell A may displace one from cell B, and this might lead to the expression of a previously dormant gene in cell B. There is, as yet, no way of guiding this exchange of genetic material between two fused cells. However, as with all microbiological experiments, one is dealing with vast numbers of cells. So long as some of the exchanges are productive this is a potentially profitable route towards the production of novel antibiotics.

### Monoclonal antibodies – 'magic bullets' at last?

The German scientist, Paul Ehrlich, secured his place in history by laying the foundations of immunology – the study of the body's defences against infection – and by investigating the effects of chemicals on microbes, which culminated at the turn of the century in the discovery of the first effective treatment for syphilis. He is almost equally well remembered for the words 'magic bullets', which encapsulated the ideal of all drug research. Ehrlich aspired to find drugs which would eradicate microbes while producing no ill effects in the patients. So far his dream has never been completely realized; even the best drugs occasionally provoke harmful side-effects. Perhaps such perfect drugs will never be found, but in 1975 scientists came a step nearer achieving something implicit in the notion of magic bullets – the ability to direct drugs to the exact position in the body where they will do most good. In that year Georges Köhler and Cesar Milstein, working in Cambridge, England, discovered how to make monoclonal antibodies.

Of all the new biotechnologies, those that utilize monoclonal antibodies are likely to have the most rapid and widespread impact on medicine. To accept this bold claim it is necessary to look at what monoclonal antibodies are and how they are made.

All animals face a constant assault from viruses, bacteria, fungi and chemicals in their environment. If these enter the body and its individual cells the results may be devastating. The first lines of defence are formed by the skin and the membranes which surround cells, and in many cases these can ward off the threat. However, when these defences are breached or circumvented, more subtle and powerful weapons are brought to bear on the invaders. Pre-eminent among the immune system's weapons are a group of proteins known as antibodies.

Antibodies are manufactured by specialized cells in the spleen, blood and lymph glands. These so-called B-cells release antibodies which roam the body, seeking out and latching on to microbes or other foreign materials. Once the invader has been tagged by antibodies, the rest of the immune system swings into action, culminating in the demise of the undesirable alien. A central puzzle of immunology – and one which has great relevance for biotechnology – is how antibodies recognize foreign substances and attach themselves to them. The answer is revealed by the molecular structure of antibodies, and this gives the clue to their enormous potential in biotechnology.

The shape of each antibody molecule is dictated by the sequence of amino acids used in its construction. All antibodies have the same basic form, that of the letter Y, but when examined in more detail it is found that this apparent uniformity disguises an almost incredible diversity. Each antibody has two identical pockets, one at the end of each arm of the molecule. The shape of these pockets varies subtly from one type of antibody to another. It is this variation which endows antibodies with their most important characteristic – specificity. The pockets of antibodies mesh with molecular structures, and hold on tightly to them (see Figure 1). In the human body there are, quite literally, millions of different types of antibodies, each having pockets with a characteristic shape.

The surface of every substance, be it a virus, a bacterium or even smooth plastic, is studded with molecules which jut out into its surroundings. When an antibody encounters a protrusion which happens to fit its pockets, the two lock together. The structure to which an antibody binds is called its antigen, and the relationship between an antibody and its antigen is very precise; an antibody will only latch on to an antigen which has exactly the right shape. The term monoclonal antibodies is applied to a group of identical antibodies all with the same shape of pocket and, thus, recognizing exactly the same antigen.

This very specific interaction can be utilized in many ways. For example, many diseases are characterized by the presence of unusual substances in the body or excessive amounts of some normal material. Monoclonal antibodies will soon be commonly employed to detect the presence of viruses, bacteria and other infections, by mixing appropriate antibodies with samples of blood or other bodily fluids. By providing a precise measure of the amount of specific substances in a patient's body, it will also be possible to diagnose many other disorders, such as some forms of infertility which are characterized by a lack of particular hormones.

The major technical difficulty lies in obtaining the required antibodies.

There is no practical way of picking out the desired antibody from the teeming hordes of other types, so another approach must be adopted. The cells that make antibodies can be separated fairly easily. This would not brighten the prospects if it were not for one fact of nature and one crucial scientific trick. The relevant fact is that each B cell (the cells that make antibodies) produces only *one* type of antibody. Could a B cell, placed in a glass dish with suitable nutrients, grow and divide to produce a clone of identical cells, all making the same kind of antibody? This might not be the required antibody, but it would be pure or monoclonal, which is more than half the battle. By repeating the process for many individual cells, persistence, technical ingenuity and luck might turn up the right antibody-producing cell eventually. Unfortunately, this neat plan founders on a simple but profound obstacle – B cells die soon after they are removed from the body. Their brief period of life in the laboratory is insufficient to provide a large enough clone even to find out exactly what kind of antibody they are producing, let alone obtain enough antibody for practical applications.

Köhler and Milstein invented the technical trick that solves the

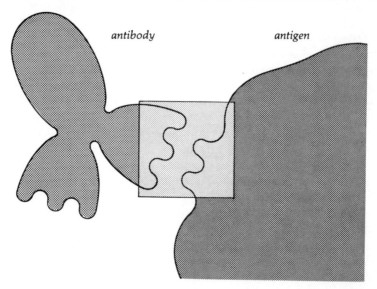

**Figure 1**
Antibodies (also called immunoglobulins) are Y-shaped proteins. At the end of each arm are two identical pockets, the shape of which varies from one antibody molecule to another. When an antibody encounters an antigen whose shape matches the pockets, the antibody and antigen bind together.

problem. Ironically this discovery, which promises so much in terms of cancer therapy, depends on the use of cancer cells. Unlike normal cells, cancer cells can be cultured quite easily. Given the right laboratory conditions they will grow and divide virtually indefinitely. Combining the antibody-producing skills of B cells with the quasi-immortality of cancer cells should produce large quantities of monoclonal antibodies. Rather surprisingly, this can be done in a very straightforward manner. Fusing a B cell with a cancer cell produces a hybrid cell, which has the properties of both original cells – it makes antibodies and it lives for a very long time. It was an arduous task to perfect a method of making these hybridoma cells, but within a few years of Milstein and Köhler's original success, the techniques had been refined and brought into common use in hundreds of laboratories.

Needless to say, these two types of cells will not join to form a hybridoma of their own accord. The biotechnologist's craft is to manipulate the cells' environment to induce them to fuse together. In the early days this was done by mixing the cells with viruses. This affects their membranes which coalesce to form a single membrane around the contents of both original cells. More recently it was found that a simple, cheap chemical, polyethylene glycol, can do the job even better.

The cells used most frequently are mouse spleen cells and mouse myeloma cells, the latter being a cancerous form of B cells. The first step is to ensure that the mouse spleen has as many B cells producing the particular required antibodies as possible. One of the most successful applications of the hybridoma technique has been to produce monoclonal antibodies which recognize interferon molecules – the natural substances that help the body fight off viral infections and may be used to treat cancer (see pages 84 and 108). To do this a mixture of materials known to contain interferon is injected into a mouse (see Figure 2). This mixture may be obtained, for example, from the blood of someone suffering from a viral disease. The substances in the mixture then act as antigens, stimulating the mouse to produce many spleen cells which make antibodies against these foreign materials, and among them will be anti-interferon antibodies. This is exactly the type of immune response which makes vaccination of human beings against infectious diseases so effective. In vaccination, a dead or severely weakened micro-organism is introduced into the body. The immune system makes antibodies against the foreign material and, having 'learned' to recognize this particular threat, the immune system is prepared to combat any subsequent invasion by a virulent organism of the same kind.

Having ensured that the mouse spleen has many anti-interferon B cells, the spleen is removed and minced. Then about fifty million myeloma cells are mixed with about double that number of spleen B cells and the polyethylene glycol is added. Only a small proportion of this huge number of cells will fuse together correctly, and of these only some

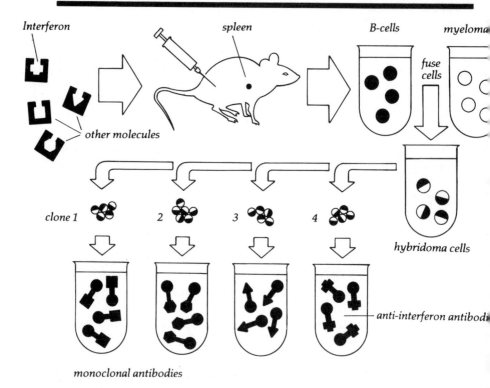

*Interferon* *spleen* *B-cells* *myeloma*

*other molecules*

*fuse cells*

*clone 1* 2 3 4

*hybridoma cells*

*anti-interferon antibody*

*monoclonal antibodies*

**Figure 2**

The production of hybridoma cells which manufacture monoclonal antibodies that recognize and latch on to interferon molecules. A mouse is injected with a mixture of materials containing a small amount of interferon. A few days after the injection the mouse's spleen is removed and its B cells, some of which will be producing antibodies which recognize interferon, are fused with cancerous myeloma cells to yield hybridomas. The hybridoma clones are separated from each other and tested to see which produces anti-interferon antibodies.

will be the required type of hybridoma. Removing the unaltered B cells and myelomas is easily accomplished; the ordinary B cells soon die, and a combination of chemicals, known by their initials, HAT, is put into the mix to kill the unaltered myelomas.

Then the considerable number of fully fledged hybridomas are grown separately to give large clones. The antibody product of each clone can be checked to see which reacts with interferon. Eventually, with a little luck, the right clone will be discovered and this can then be used immediately to make large amounts of antibody or frozen until it is wanted. Compare

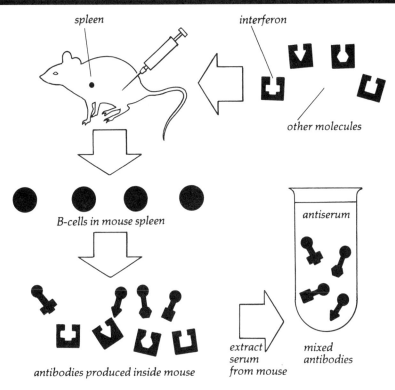

**Figure 3**
The production of conventional antisera consisting of a mixed population of antibodies. Antisera have been used for many years in diagnosis and, sometimes, therapy. A mouse (or often a goat or a horse) is injected with material that contains the substance which antibodies are needed against. Blood taken from the animal contains a certain amount of the desired antibody, mixed with many other unwanted materials. Since conventional antisera are obtained from live animals and are impure, the process is both costly and has many limitations.

this end result with the mixed population of antibodies present in conventional antisera (see Figure 3).

In theory, and increasingly in practice also, biotechnologists can develop monoclonal antibodies which are specific for virtually any substance of major medical interest. The pages of innumerable scientific journals are filled with details of experiments which employ monoclonal antibodies in dozens of ways – from identifying cancer cells to tracing the nerves of giant leeches. In the last five years these new tools have played an ever more important part in basic research. To meet the demand for monoclonal antibodies, exotically named companies, such as Hybridtech, Celltech and Atlantic Antibodies Inc., sprang up like mushrooms in the late seventies and early eighties.

This new technology is now poised to make a major impact in medical practice over the entire spectrum of human diseases, from cancer and kidney failure to infections caused by viruses and bacteria. The potential applications of monoclonal antibodies include boosting the natural defences of patients; improving the chances of successful organ transplantations; targeting drugs to specific parts of the body; and purifying drugs before they are administered to patients.

An example of this final group of applications, which has already been described, is the production of monoclonal antibodies which attach themselves only to interferon molecules, spurning all others. These antibodies can be fixed on to beads which are then packed inside a tube. When a mixture of materials containing a tiny percentage of interferon is poured into the tube, the antibodies pluck the interferon molecules from the soup while the unwanted materials flow away. Chemicals are then passed through the tube to loosen the antibodies' hold on interferon which is flushed out. This technique yields a solution which is 5000 times richer in interferon than the original crude mixture. This and other uses of monoclonal antibodies rely on their exquisite powers of discrimination. Looking for a needle in a haystack, and extracting it, is not such a problem if one has a magnet – and in monoclonal antibodies we can hope to find a 'magnet' which will pick up just what we are looking for.

### Interferon – the natural anti-viral alarm
The chequered history of interferon began in 1957 in London, where Alick Isaacs and Jean Lindenmann were investigating the body's responses to viral infections. They found that a substance secreted from infected cells helped other cells resist the effects of invading viruses. They named this substance interferon because it appeared to interfere with the spread of viral infections. Since then this enigmatic material has shuttled from prominence to obscurity and back again. Until recently, all attempts to elucidate the true nature and role of interferon foundered due to a lack of sufficient pure material for investigation. Now, interferon has caught the imagination of the newly emerged band of genetic engineers who have been able to provide larger quantities of interferon, which can be purified with the aid of monoclonal antibodies. Thus the way is now clear for a detailed evaluation of interferon's potential for the treatment of viral diseases and some cancers.

Isaacs and Lindenmann were trying to find out why people suffering from an infection by one kind of virus rarely caught another type of viral disease at the same time. This fact intrigued them, for it was well known that one kind of bacterial infection often paves the way for another because the patient's defences are weakened. Why should viral infections be different? Isaacs and Lindenmann provided the clue when they discovered that, in response to an invasion by viruses, cells secrete interferon which acts as an alarm signal, alerting neighbouring cells to the invaders' presence and allowing them to prepare themselves against the imminent attack. It is now known that there are at least a dozen different kinds of interferon produced by human cells, all of which are proteins. Furthermore, the interferons produced by different species of animals are themselves different; interferon from mice has little effect on humans.

Until 1980 the sole source of human interferon was, not surprisingly, human cells. Most of the world's supply came from the laboratory of Kari Cantell in Helsinki, where vast quantities of white blood cells from donors were deliberately infected with viruses, stimulating them to produce interferon. The tiny amount of interferon manufactured by each cell and the difficulty of separating it from all the other materials meant that the blood from 90,000 donors provided only 1g of interferon in a form which was, at best, only 1 per cent pure. The scarcity of interferon and its enormous price, sometimes estimated at $50 million per g, constituted a severe brake on research. (The prices calculated for interferon were, in fact, rather specious since no-one ever possessed as much as a gram at any one time, let alone offered it for sale in such quantities. Nevertheless, the cost was extraordinarily high.)

Despite these difficulties, evidence for the value of interferon in the treatment of some viral diseases and, perhaps, some cancers did begin to accumulate. Then in January 1980 news from Switzerland sparked the interferon boom, both clinically and financially. Charles Weissmann announced that he and his colleagues in Zurich had cloned the gene for one type of human interferon and inserted it into bacteria which promptly began manufacturing interferon. Other researchers soon followed suit and, within a year or so, the world's supply of interferon had increased dramatically, its cost had plummeted by about 90 per cent and, most importantly, very pure interferon became available at last.

By 1983, about thirty companies in the US alone had become involved in the interferon business and many countries had instigated clinical trials. The conviction that interferon merits very serious consideration has spread far beyond the small band of early enthusiasts. Interferon has been hailed as the new hope for cancer sufferers and, indeed, there are hopeful signs that it could come to play a part in cancer therapy (p. 108). Much closer at hand is the use of interferon to prevent or cure several viral diseases.

Preliminary studies have revealed promising results in the treatment of rabies, hepatitis, shingles, various herpes infections and cytomegalovirus infections. The last of these is a very common virus which infects about half of the population at one time or another. It can produce fever, pneumonia and even death in new-born babies and already weakened adults. Experiments with monkeys have shown that interferon can also give protection against a lethal virus which attacks the heart muscles.

The best evidence for interferon concerns its use as a prophylactic rather than a cure. Researchers at Britain's Common Cold Research Unit have shown that interferon, given as a nasal spray, builds resistance to this all too familiar illness.

At present the cost of interferon is still far too high for it to be used against relatively trivial infections such as the common cold, even though it causes millions of lost working days. Many biotechnologists believe, however, that improved manufacturing methods may bring its price down dramatically and that it may be sold over the chemist's counter. In the USSR 'interferon' is already on sale to the public, but the dosages are so low that the preparations are almost certainly ineffective.

So far only the first few chapters of the interferon story have been written and, undoubtedly, the plot will take many twists and turns before the full potential of these intriguing substances is revealed. The challenges for the next few years are to find out more about the way they work, which types of interferon are best for each disease, and what harmful side-effects they produce. These are tasks for basic scientists and clinicians. Biotechnologists are trying to develop more effective and cheaper methods of supplying pure interferon for research and, eventually, for its widespread use in medicine.

### Anti-viral vaccines – protecting the masses

In 1967 over 10 million people were infected with smallpox and the disease was endemic in more than thirty countries. Today, this hideous affliction has been wiped off the map. The World Health Organization's mass vaccination programme against smallpox is arguably the greatest triumph of modern medicine, and its success highlights the immense benefits which can accrue from the development of effective vaccines against viral diseases. Polio, yellow fever, rabies and rubella (German measles) are just a few of the other viral diseases which can now be fought with vaccines. There remain, however, many widespread and serious viral diseases for which no cheap and effective vaccine exists, and biotechnologists are in the forefront of the crusade to bring these intractable diseases under control.

The US Office of Technology Assessment points to seven major viral diseases in humans for which vaccines are needed: hepatitis, influenza, herpes simplex, mumps, measles, common cold and varicella-zoster (shingles). The encouraging progress being made towards a vaccine against hepatitis illustrates many of the general themes of the biotechnological approach to viral vaccine production.

Three major forms of hepatitis exist: two are termed hepatitis A and B, and the third is known, rather unhelpfully, as non-A non-B hepatitis. Most attention has focused on hepatitis B, serum hepatitis. The virus that causes this disease has been identified and is responsible for thousands of serious liver infections and roughly 100 deaths a year in Britain, while in the US, the Center for Disease Control estimates that there may be as many as 150,000 cases a year.

The standard method of producing anti-viral vaccines is to grow the virus either inside a suitable animal or, preferably, inside cells grown in the laboratory. The viruses are then collected and either killed or severely weakened before being injected into humans. In response to these foreign intruders, the body's immune system makes antibodies which will attack them. This takes time, but since the viruses are in a harmless form the delay does not matter; the dead or feeble viruses will do no damage while the immune system is gathering its strength. If, later, a live and fully active virus of the same type gains access to the body, the immune system is already prepared to stamp it out.

The crux of the problem is that there is often no convenient method for growing large quantities of a particular virus. Viruses are amazingly simple structures consisting of just a small piece of genetic material encased in a coat of protein. Peter Medawar, an eminent British immunologist, has summed up viruses as 'just bad news wrapped in protein'. Fortunately, as explained in the discussion of monoclonal antibodies, the immune system only recognizes foreign substances by their external features, in this case the viral protein coat. This fact is being used by genetic engineers as part of a plan to produce an anti-hepatitis vaccine. They have taken the genes that code for part of the hepatitis coat protein and introduced them into the bacterium, E. coli. The bacteria manufacture the viral protein without, of course, making any of the viruses themselves. If all goes well with the clinical trials now under way, this genetically engineered protein will be available as a vaccine within a year or two.

This general approach can be applied to the prevention of the spectrum of viral diseases. Apart from the human diseases mentioned earlier, some of the most rapid progress has been made in the veterinary area, most notably with a vaccine for foot-and-mouth disease (see p. 129). The first stage is to identify a characteristic feature of the particular virus' protein coat, and the exquisite powers of discrimination offered by monoclonal antibodies are invaluable here. Genetic engineers then clone the appropriate gene and persuade bacteria to synthesize it in large quantities. Among recent advances in this area is the cloning of a rabies virus antigen by scientists at the French firm of Transgene and several groups of researchers have turned their attention to the viruses that cause influenza, the common cold and herpes.

Genetic engineering not only promises to provide protection against diseases that cannot be fought by conventional vaccines, it should also mean safer versions of existing anti-viral vaccines. Viruses themselves will not be employed in these new vaccines but only one of their proteins made by bacteria, eliminating the risk that the vaccine could be contaminated with live viruses. This is something that happens extremely rarely now, but expensive and time-consuming procedures are needed to ensure that such an accident does not occur.

### Parasites – the enemies within

Over 1000 million people in the tropics live under the threat of painful and mutilating parasitic diseases which exact a toll of millions of human deaths each year. Even more distressing is the fact that in many areas these diseases are becoming *more*, not less, common. The greatest benefit biotechnology could bring to humans would be to help lift the scourge of these diseases.

There is a clear possibility that genetic engineering will produce new and effective vaccines for many parasitic diseases, but in the near future monoclonal antibodies are likely to be of much greater importance. A major problem that faces all doctors, especially those working in ill-equipped hospitals and clinics, is to make accurate diagnoses of diseases. A battery of individual monoclonal antibody preparations which can be used to identify specific parasites would be of immense benefit (see Figure 4). The production of monoclonal antibodies which recognize characteristic features of particular parasites is now beginning, and their use in simple tests should become possible within a very few years.

In 1975, the United Nations Development Programme, the World Bank and the World Health Organization set up a Special Programme for Research and Training in Tropical Diseases. Six diseases were targeted for a concerted attack, two of which, malaria and leprosy, are well known to everyone. While the other four, filariasis, leishmaniasis, schistosomiasis and trypanosomiasis, may be less familiar, the threat they pose is just as great. The efforts of biotechnologists are showing signs of success in the treatment of three of these diseases, malaria, leprosy and leishmaniasis.

### Malaria

In the nineteen-fifties it seemed that malaria would soon be conquered – perhaps not eradicated entirely, but certainly stamped out as a major cause of death in many countries. By the seventies, such confidence had evaporated. Malaria had fought back against all that science and medicine could throw at it and the disease was re-establishing itself in 'clean' areas.

The optimism was engendered by the initial success of a two-pronged attack on malaria: the malarial parasite itself was being killed with drugs; and DDT and other insecticides were controlling the mosquitos which spread the disease from person to person. Increasingly, however, both the parasites and the mosquitos have become resistant to the chemicals which previously killed them with ease. A new approach is urgently required, and the production of vaccines with the aid of biotechnology is now a real possibility.

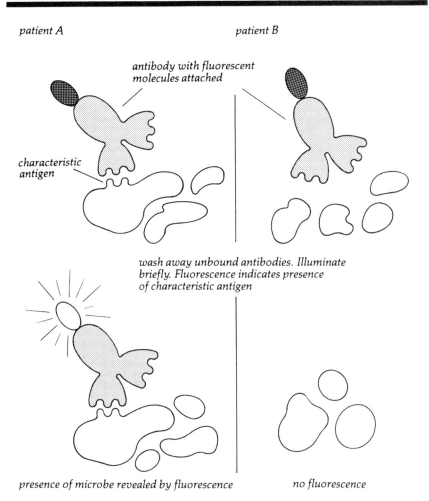

patient A

patient B

antibody with fluorescent molecules attached

characteristic antigen

wash away unbound antibodies. Illuminate briefly. Fluorescence indicates presence of characteristic antigen

presence of microbe revealed by fluorescence

no fluorescence

## Figure 4

Monoclonal antibodies in diagnosis. The presence of specific materials in samples taken from a patient will help doctors diagnose an illness. Microbes, viruses, proteins and many other substances all have characteristic antigens. Monoclonal antibodies which recognize specific antigens can be linked to fluorescent molecules which shine under ultraviolet light. These fluorescent antibodies are added to a sample of blood or other material taken from the patient. If the antigen is present, the antibodies will cling to it. The sample is then washed to remove unbound antibodies. Any light emitted after the washing process betrays the presence of the particular antigen for which the test has been devised. Fluorescent markers are just one of several ways of making visible the presence of antibodies. Radioactive labels can also be employed.

The chills, fever and profuse sweating so characteristic of malarial infections are the body's responses to millions of tiny, single-celled organisms which have invaded the patient's red blood cells. The malarial parasites, *Plasmodia*, are not bacteria but belong to the group of organisms known as protozoans. Like our own cells, they have well-defined nuclei enveloping their DNA and are unaffected by the antibiotics which kill bacteria.

The malarial parasites are transferred from one person to another by female mosquitos which feed on human blood. It is quite impossible to catch malaria directly from an infected person. *Plasmodia*, injected into the bloodstream by the mosquito's bite, migrate to the liver where they proliferate. Within a few days as many as 30,000 microbes burst out of each liver cell and invade red blood cells. There the *Plasmodia* grow and multiply until eventually they burst out in search of fresh cells to infect. Strangely, the millions of individual *Plasmodia* act in concert; they all stream out of infected cells at about the same time, repeating the process every two or three days. This phenomenon is reflected in the cyclical nature of malarial fever, in which periods of relative calm are punctuated with acute attacks.

One of the reasons for the malarial parasite's 'success' is that while inside blood cells it is hidden from the patient's immune system. However, it is exposed to danger during the brief movement from one cell to another. All too obviously, the human immune system often cannot meet the challenge. A good vaccine which 'teaches' the body how to make antibodies against malarial parasites before the infection occurs would make all the difference. People in high-risk areas would then be able to fight off the invading parasites as soon as they enter, while there are only a few to deal with. Without vaccination, the inevitable delay in mounting an immune defence gives the parasites time to build up their forces.

It was not until 1976 that William Trager, at New York's Rockefeller University, discovered how to grow *Plasmodia* in the laboratory. At first it appeared that this might lead to the production of anti-malarial vaccines. The standard method of growing large amounts of the parasites, killing them and injecting the remains into humans, was tried. Although initial tests on animals were encouraging a major, and possibly insuperable, obstacle intervened. The parasites were grown in the laboratory inside human blood cells. Unless all traces of the human cells are removed from the vaccine the body is likely to react against these as well as the parasites, and the results of this response could be dire.

Recently scientists – notably Ruth Nussenzweig and her colleagues at New York University – have made monoclonal antibodies which recognize a protein on the surface of *Plasmodia*. Mice and monkeys injected with these monoclonal antibodies resisted malarial infections. It is conceivable that this could form the basis of a passive immunization programme. Such immunization is termed passive because the protection depends on foreign antibodies injected into the body, rather than on antibodies made by the individual's own immune system, as in active immunization.

Far more exciting, however, is the possibility that large amounts of the parasite's protein could be made with the aid of genetic engineering. Pure samples of this protein could then be inoculated into human beings, making them immune to the malarial parasite.

## Trypanosomes – many-faced menaces

The term trypanosomiasis covers several intractable, tropical diseases, each caused by a different species of the group of protozoans called trypanosomes. The two commonest forms of these diseases in humans are Chagas' disease, which afflicts 10 million people in South America, and African sleeping sickness, which claims 10,000 fresh victims each year. In Chagas' disease, *Trypanosoma cruzi* invades the heart, nervous system and gut, and it is especially fatal in children. The two species of trypanosomes responsible for sleeping sickness also attack the nervous system and, unless treated, the disease is invariably fatal. The indirect effect of trypanosomes on human welfare is also great. Other species of trypanosomes infect cattle (killing three million a year), sheep and goats, preventing the inhabitants of vast tracts of Africa from rearing these animals for food.

It must be stated at the outset that biotechnology probably has little to offer in the short term for the alleviation of this suffering. Nevertheless, there is one contribution biotechnologists might make, and a brief look at trypanosomes will also emphasize the huge challenge faced in attempting to overcome the natural cunning of parasitic organisms.

The work of immunologists, and, more recently, molecular biologists studying the genes of these organisms, has revealed a truly remarkable strategy for avoiding the attentions of the human immune system. Trypanosomes are masters of disguise. The antigens on the surface of most cells are relatively constant – those found one day will probably still be there the next. Trypanosomes are quite different; they can call upon a well-stocked 'wardrobe' of antigens with which to dress themselves. No sooner has the patient's immune system learned to recognize a trypanosome by one of its antigenic outfits, than the trypanosome dons another set of antigens plucked from its store. The whole process of identifying the intruder must start anew, and so on almost indefinitely. Consequently, the immune system is never able to mount an effective campaign against these chameleon-like organisms.

This behaviour results from the trypanosomes' ability to switch on and off an array of different genes, each coding for different antigens. It rules out any reasonable hope of producing vaccines against the disease, for dozens of separate vaccines would be needed, one for each of the organism's many guises. Perhaps the only good thing that can be said for these detestable creatures is that through studying them a great deal has been learned about the way genes can be manipulated, and this knowledge may yet underpin new applications of genetic engineering.

However, biotechnology might help combat sleeping sickness by the production of two types of protein. Recently an anti-cancer drug, daunorubicin, was found to be active against trypanosomes if the drug is first linked to albumin or ferritin, two proteins found in human blood. The supply of these proteins is limited at present. If they were to form part of a treatment for sleeping sickness, biotechnologists would undoubtedly be called upon to produce the required amounts, probably by inserting the appropriate genes into bacteria and purifying the protein thus manufactured.

## Leprosy – myths, misunderstandings and modern medicine

Other diseases may have attracted more folk tales, but few are as well-entrenched as those surrounding leprosy. One is that leprosy is highly contagious; in fact, about 95 per cent of the population is probably immune to the disease and even susceptible individuals will only catch it after long-term exposure to the infection. Another myth claims, wrongly, that leprosy causes parts of the body to drop off. It does, however, produce numbness in the extremities of the body, making accidental damage more likely. Finally, while it is true that the overwhelming majority of lepers live in tropical countries, cases of the disease are by no means unknown elsewhere. For example, there are over 4000 cases in the US among recent immigrants, and only a century ago leprosy was still found in northern Norway.

Leprosy is caused by a bacterium called *Mycobacterium leprae*, and about 11 million people are affected by the disease. Until recently the chemical dapsone formed the basis of an effective treatment, but the alarming rise in bacteria resistant to dapsone demands new types of therapy, at much greater cost to the poor nations where leprosy is common.

One of the principal aims of the Special Programme mentioned earlier is to develop reliable skin tests to identify people at particular risk from the disease. Very few animals can become infected with leprosy bacteria, a surprising exception being the nine-banded armadillo. Purified bacteria from armadillos are being used in small-scale tests, but this may be impractical for general use. Obviously a vaccine is needed, but there is, as yet, no way of growing the bacteria in the laboratory so that a vaccine can be manufactured. Unfortunately, there seems to be little research devoted towards this important challenge.

## Liposomes – lethal packages for leishmania parasites

An entirely new approach towards more efficient drug delivery is now proving its worth in the treatment of one of the most dangerous parasitic diseases, leishmaniasis, and is rapidly becoming a prominent feature of many areas of biotechnology. Tiny globules of fat-like material, known as liposomes, can deliver potent poisons to cells infected with leishmania parasites. Since many millions of people are afflicted with these microbes, liposomes could clearly make a great contribution to human welfare.

The membranes which form such a vital part of all cells contain large amounts of materials called phospholipids. The many different types of phospholipids share the same basic structure – two long 'tails' composed of fatty materials and a 'head' containing nitrogen, oxygen and phosphorus atoms. Most people have seen how fatty or oily liquids tend to accumulate into droplets when poured on water, rather than spreading evenly across the surface. Substances which behave in this manner are called hydrophobic (water-fearing). The fatty tails of phospholipids (the lipid parts) are hydrophobic and shun contact with water but the phospholipid heads avidly seek out the company of water molecules and so are hydrophilic. Figure 5 illustrates one very efficient way in which these two opposing inclinations of a phospholipid molecule can be reconciled.

In the nineteen-sixties, scientists capitalized on this natural phenomenon by learning how to assemble phospholipids into hollow spheres called liposomes. Some consist of just one sphere, while others are made up of several phospholipid shells, one inside the other rather like Russian

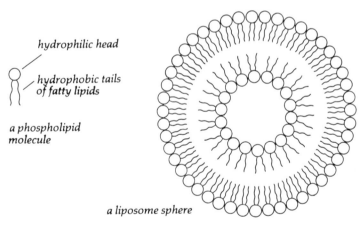

hydrophilic head

hydrophobic tails
of fatty lipids

a phospholipid
molecule

a liposome sphere

**Figure 5**
Phospholipids and liposomes. A phospholipid molecule (left) consists of two hydrophobic 'tails' of lipids, and a hydrophilic 'head'. When large numbers of phospholipids are mixed with water they assemble themselves into spherical structures called liposomes. The cross-section of a liposome (right) shows two concentric spheres of phospholipids with a water-filled space in the centre.

dolls. The space in the middle can be used to entrap all manner of materials, including drugs. In the last few years there has been much research into the advantages of wrapping drugs inside liposome packages, and the treatment of leishmaniasis is one of the most promising applications.

Leishmaniasis afflicts roughly 100 million people in South and Central America, Africa, Asia and parts of Southern Europe. The disease takes several forms, each associated with a different species of protozoans called *Leishmania*. The most devastating is kala azar in which *Leishmania donovani* invades the spleen, liver and bone marrow, causing almost certain death unless properly treated. An epidemic of kala azar spread among over 70,000 people in the Indian State of Bihar in the late seventies, killing more than 4000.

There is no effective vaccine against kala azar, and the sick are usually treated with compounds which contain antimony. Antimony compounds are similar to arsenic compounds, and so it is not surprising that there is a very real danger of poisoning the patients rather than curing them. Although modern antimony drugs are safer than those used some years ago, the basic problem remains – antimony does human cells no good at all.

Liposomes offer a major advance in terms of safety, for antimony drugs are up to 700 times more effective if they are first encapsulated inside liposomes. This does not mean that the patients are cured that much more rapidly, rather that only a fraction of the normal dose of the drug need be given to the patients, greatly reducing the risks. This trick seems to work because liposomes injected into the bloodstream tend to be most rapidly absorbed by the same types of cells that *Leishmania* parasites colonize. Thus, liposomes laden with antimony drugs deposit them predominantly in the liver and spleen where they are needed.

Obviously this method of delivering drugs might be extended to other diseases which primarily affect these organs, and current research indicates that this is a real possibility. However, the potential of liposomes does not end here. In this example liposomes tended to home in on liver and spleen cells only, and this would appear to limit their use. Fortunately it does seem possible to make liposomes which are targeted against different types of cell. To some extent this can be achieved by altering the composition of the liposomes themselves, selecting different sorts of phospholipids for example.

It is most unlikely that the specificity of liposomes can ever approach that offered by antibodies. It turns out, however, that these two entirely distinct tools can be combined, perhaps giving the best of both worlds. Liposomes loaded with drugs can be linked to antibodies which recognize specific microbes, and so the antibodies provide the guidance system, while the liposomes carry the 'warheads'.

Liposomes are particularly well suited to transporting substances about the body. They have been used in experiments to introduce a wide range of materials into animals, including antibiotics, enzymes, insulin and even compounds which mop up plutonium. Perhaps their most important feature is their ability to protect drugs and other materials from destruction by the hostile environment of the body. The liposome shell prevents the body's enzymes from attacking the drugs before they can do their work. Furthermore, liposomes are non-toxic and do not usually provoke a response from the immune system.

# Correcting nature's errors

As well as combating diseases caused by viruses, bacteria and protozoans, biotechnology can also help cure disorders which result from imbalances or defects in the body's chemistry, including some inherited diseases. It is now possible to manufacture an enormous range of natural compounds which can or might relieve illnesses such as diabetes, arthritis, bone disorders and nerve damage, as well as producing pain killers and materials to treat burns and wounds.

### Hormones – the body's messengers

Our bodies contain a dazzling array of different hormones, each with a vital task in coordinating the activities of individual cells and tissues. Without hormones, cells would have little idea of the overall state of the body and could not know what tasks were required of them to keep the whole organism functioning properly. Hormones convey all sorts of information – for example, that food is available for digestion; that danger threatens and muscle cells must prepare for action; and that a child is developing in the womb.

Not surprisingly, a defect in one of the hormone systems can produce serious illnesses. These can sometimes be remedied by supplying the patient with the right amount of a particular hormone, and biotechnology can help in their manufacture.

Many of the body's hormones are composed of chains of amino acids. The term protein is generally reserved for chains of about twenty or more amino acids and some hormones have very short chains. Therefore, the term polypeptide hormone is usually employed for all hormones made up of amino acids (after the chemical link that joins two adjacent amino acids, called a peptide bond). The best-known polypeptide hormone is insulin and a great deal is understood about how it maintains the correct amount of sugar in the bloodstream, and how some types of diabetes can be treated by supplying patients with extra insulin. However, there are many other important polypeptide hormones, the names of which are likely to become more familiar as they come to be widely used in medicine through the efforts of the genetic engineers who will help increase their supply. They include growth hormone, a deficiency of which causes some forms of dwarfism, and nerve growth factor, which could help in the treatment of injuries and restore nerves after surgery.

Genetic engineers are turning their attention to three categories of substances with proven or possible medical applications, and polypeptide hormones provide examples of all these. First, there are substances for which we already have a source of supply, but cheaper and better materials might be made by microbes (for example, insulin). Second, there are substances with a known medical value whose supply is inadequate to meet the present demand (for example, growth hormone). Finally, there are substances which may prove to be very useful, but larger quantities are needed before their effects can be tested (for example, nerve growth factor).

Until the advent of genetic engineering all the insulin given to diabetics was extracted from the pancreas of cattle or pigs. It is estimated that there are 60 million diabetics in the world, more than half of whom live in developing countries where they are rarely diagnosed or treated. About one third of the approximately 15 million known diabetics in the developed world are currently receiving an average of 2mg of insulin a day. Insulin derived from animals costs about $75 per kg to produce, but one of the world's leading manufacturers estimates that this might be reduced by two-thirds if it could be made on a large scale by genetically engineered microbes. (This dramatic decrease in the cost of producing insulin would not, however, be reflected in an equivalent reduction in the cost of the drug to the patient, since many other factors contribute to this, such as packaging and distribution, which would remain unchanged.)

The intense interest in genetically engineered insulin rests largely on two factors. First, insulin is a test case of whether genetically engineered microbes can be efficiently incorporated into the pharmaceutical industry. Second, there are hopes that this insulin may be better and safer than animal insulins. The insulin molecules made by cattle and pigs are very slightly different from those made by humans. The insulin molecule consists of fifty-one amino acids, and at one point the pig insulin has a different amino acid compared to the human version, while cattle insulin differs at three points. These small variations mean that some patients eventually suffer allergic reactions to the animal hormones. Insulin produced by microbes is exactly the same as human insulin. It will take time to discover whether this makes a substantial difference in practice, but theoretically it should bring a benefit to some diabetics. Human insulin may also prevent or reduce some of the long-term effects of diabetes, which include damage to the retina leading to blindness and kidney damage. The production of human insulin by genetic engineering is discussed in more detail in Chapter 2.

The major problem in the case of growth hormone is lack of supply. A considerable proportion of children who fail to grow to a normal height

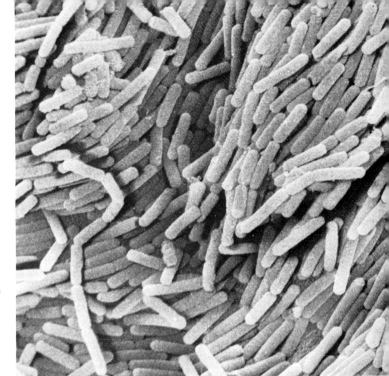

**1** (*right*) Rod-shaped bacteria (*Bacillus cerus*) magnified 1250 times

**2** (*below*) Thread-like molecules of DNA spilling out of a damaged bacterium

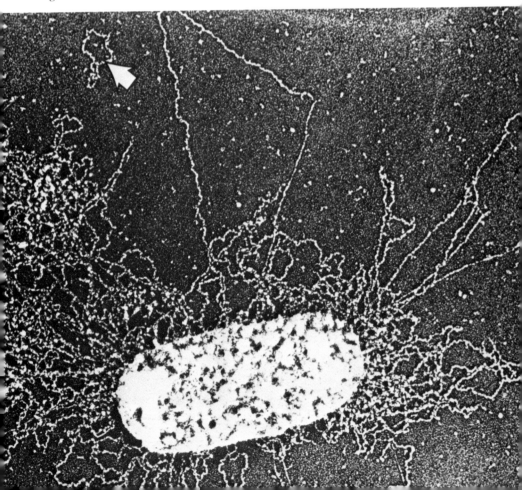

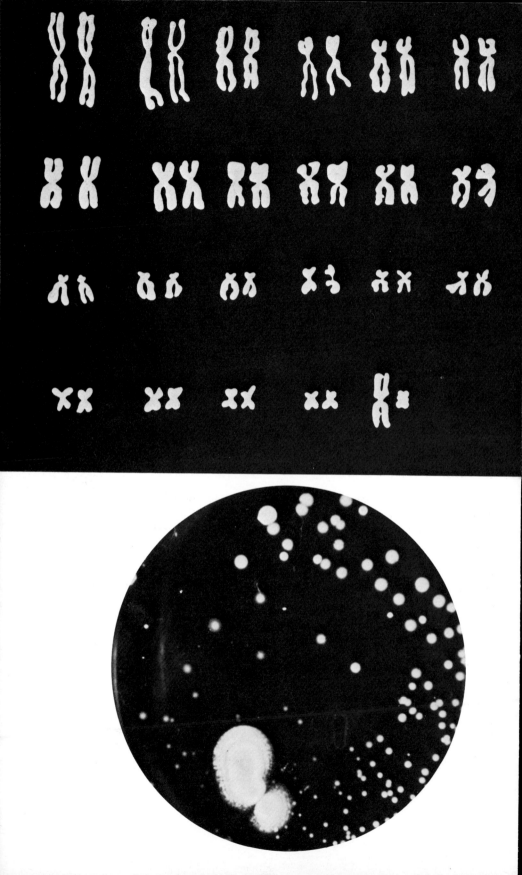

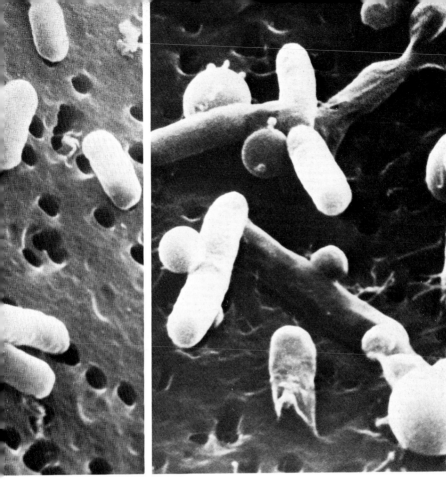

**3** (*above left*) A set of human male chromosomes. Humans have 23 pairs of chromosomes and between them they contain all the genetic information, encoded in DNA, required for life. Bacteria have only one, circular, chromosome.

**4** (*left*) The effect of *Penicillium* mould on the growth of bacteria. One of the stock pieces of equipment used by microbiologists is a round glass dish (a Petri dish) which can be filled with a jelly-like material on which microbes thrive. The Petri dish in this illustration has been used to reproduce Fleming's original observation. The *Penicillium* mould on the left is releasing penicillin into the surroundings and this has stopped the bacteria growing near it. The light blobs on the left are colonies of bacteria growing untouched by the penicillin.

**5** (*above*) The action of penicillin. Penicillin interferes with the bacterium's ability to build up its cell walls correctly. On the left are normal cells of the bacterium *Escherichia coli*. On the right are bacteria treated with penicillin. The cells are swelling and bursting because they have not been able to manufacture strong cell walls. Penicillin has little or no effect on animal cells because the membrane that surrounds these cells is made of different chemical compounds.

**6** Penicillin is produced in huge metal vats with capacities of 100,000 litres or more. The *Penicillium chrysogenum* mould is provided with an ideal environment and supplied with nutrients, oxygen and other necessary substances via a complicated set of pipes and valves. Penicillin fermentations take two to three weeks to complete. Pharmaceutical factories may, therefore, have a dozen or more fermenters so that there is a regular supply of penicillin mixture for purification and further processing.

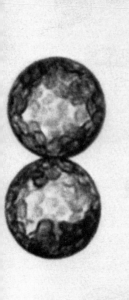

7 (*above*) Two tobacco protoplasts (cells with their cell walls removed) in the process of fusing to form a single cell

8 (*right*) Crown galls, cancer-like growths on plants caused by *Agrobacterium tumefaciens*. The Ti (tumour-inducing) plasmid of this bacteria inserts itself into plant cells and causes them to grow and divide in an uncontrolled way.

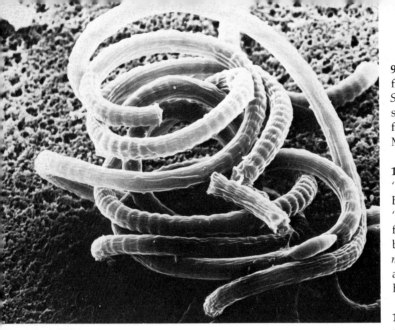

**9** (*left*) The spiral filaments of the alga *Spirulina platensis*, a rich source of protein used as food for many centuries Mexico and Chad.

**10** (*below left*) ICI's 'Pruteen' plant in Billingham, Cleveland. 'Pruteen' is an animal feed, made from the bacterium *Methylophilus methylotrophus*. The cells are grown in the 200 foot high fermenter.

**11** (*above right*) Ethanol-powered motor car in a sugar cane field in Brazil Ethanol fuel is derived from the fermentation of sugar cane.

**12** (*right*) *Zymomonas mobilis*, a bacterium that ferments sugars to alcoh seen here fixed to cotton fibres. In a 'continuous' fermentation system, in which raw materials are steadily fed in at one end and the product drawn at the other, fixing the c to a solid support prevents them from bei 'flushed out' of the syste as the raw materials flow past.

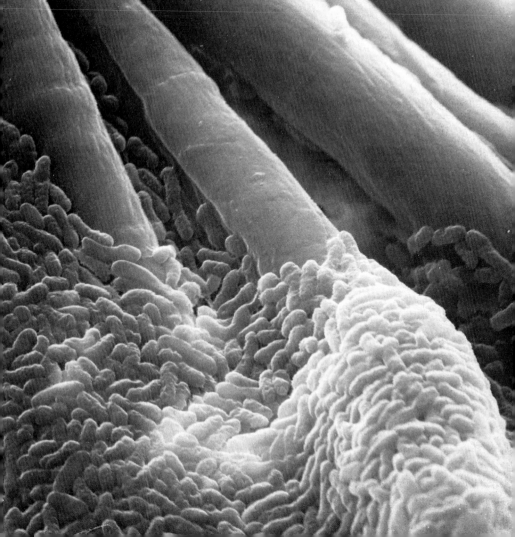

**13** Leaching operations outside Kennecott's Bingham Canyon open-pit mine near Salt Lake City, Utah. The pit is half a mile deep, and has produced approximately 12 million tonnes of copper.

do so because their pituitary glands fail to produce enough growth hormone. If the lack is diagnosed early enough, such children can be given extra growth hormone extracted from cadavers. The therapy involves an injection of about 7mg of hormone a week for several years, while a single pituitary gland provides only about 4mg. In 1981, 800 children in the UK could have benefited from such treatment, but there was not enough of the hormone available to give them all as much as they needed. In the US, about 1600 children are receiving growth hormone.

Human growth hormone produced by genetically engineered bacteria is now on trial in several countries and the indications are that it is effective. Genentech in California and Kabi-Vitrum in Sweden hope to start commercial production in 1984. A single 500 litre (110 gal.) vessel filled with these bacteria can provide the equivalent of no less than 35,000 human pituitary glands. Once growth hormone is available in relatively large quantities it can be tested for its effects on other disorders, including senile osteoporosis (a loss of calcium from bones), some bleeding ulcers, and to aid healing after burns, wounds and bone fractures.

The final group of polypeptide hormones contains less familiar but equally vital substances, including somatostatin, calcitonin, cholecystokinin, bombesin, vasopressin, parathyroid hormone, nerve growth factor, adrenocorticotropic hormone and erythropoietin. Some are well understood and for these there is already a clear medical need for better supplies to treat specific diseases. Others are more enigmatic. Only when they are all available in larger quantities, in a pure form, will their true significance be revealed and their potential medical applications become apparent. The medical and scientific communities are now looking to genetic engineers to provide these materials.

Considerable progress has already been made in some cases. For instance, somatostatin, a hormone found in the pancreas, the brain and elsewhere, was the first human molecule to be obtained from bacteria through genetic engineering – by a team of researchers in California in 1977.

At present these hormones are obtained from four sources: animals (in which case they may not be identical with the human hormones); human organs, blood or urine; purely chemical manufacturing processes; or cells grown in laboratories (cell or tissue culture). Each method has its disadvantages, for example, cost, limited supply, purity. Genetic engineering, while not without its own problems, is highly likely to prove better than any other technique for making significant quantities of at least some of these materials. The following is a brief survey of some of the present and potential medical applications.

Bone disorders affect large numbers of people – up to 3 per cent of Western Europeans over forty, according to one estimate. Among the most important factors in keeping bones healthy is the maintenance of the correct concentration of calcium in the body. Calcitonin is one of several hormones whose job is to keep the calcium balance right, and in certain types of bone disease extra calcitonin can be beneficial. Calcitonin consists of thirty-two linked amino acids and can be manufactured by chemical means. It is, in fact, the largest polypeptide manufactured in this way for pharmaceutical use. This chemical production method might be superseded if a biotechnological process, involving genetic engineering, can be developed which makes either a cheaper or a purer product. Parathyroid hormone may also find a role in the treatment of bone disorders, either on its own or in combination with calcitonin. These are two of several hormones synthesized in the thyroid gland.

By contrast, adrenocorticotropic hormone (ACTH), a slightly larger polypeptide of thirty-nine amino acids, is not manufactured chemically. ACTH is produced in the pituitary gland and is one of the most important hormones, controlling the actions of many others. It is obtained from animals and can be used to treat inflammations.

Cholecystokinin and bombesin both reduce the appetite. While the ideal solution for most overweight people is to eat less without the aid of medicaments, these two hormones would doubtless find a ready market if manufactured commercially.

Erythropoietin controls the development of blood cells and it could possibly be used to treat burns, bleeding and other conditions involving the blood.

At a more speculative level, there is some evidence that a very small polypeptide (only seven amino acids long) called MSH/ACTH 4–10 affects concentration and memory. If this is the case *and* there are no serious side-effects, then great interest would be shown in this material.

### Endorphins – nature's pain-killers

'Among the remedies which it has pleased Almighty God to give to man to relieve his sufferings, none is so universal and so efficacious as opium.' Thomas Sydenham, writing in 1680, might have added that few drugs are also so dangerous. Morphine, the major active ingredient in opium, and heroin, its close relative, are two of the most potent pain-killers known. Why should these and similar chemicals derived from poppy seeds produce such profound effects on the human body? The key to the enigma began to be revealed in 1975 when John Hughes, Hans Kosterlitz and their colleagues in Britain, discovered that two small chemicals inside pigs' brains had an effect very like opiate drugs. This initiated a flurry of research which may lead to the production of highly effective and safe pain-killers. Subsequent work in Britain, the US and elsewhere showed that there are a number of other hormones which mimic the effects of morphine and similar drugs. First to be discovered were the enkephalins, each of which consists of only five linked amino acids. Then came the endorphins, rather larger molecules, which were given their name because they are endogenous (growing from within), morphine-like substances. One of the endorphins, dynorphin, is 1200 times more potent than morphine.

The enkephalins and endorphins undoubtedly have many other effects on the body apart from deadening pain. The interaction between them is complex and subtle, but knowledge about them is accumulating rapidly. In theory, these substances should be safer than drugs manufactured chemically or extracted from plants, and this has inspired many scientists to investigate them more thoroughly. Already at least one firm, Endorphin Inc. of Seattle, is looking at the medical potential of these substances and they have contracted researchers at University College, London, to clone the gene for pancreatic endorphin. Most interest has centred on the endorphins made by the pituitary gland, but it seems that, when injected, these endorphins do not easily enter the brain in an active form, whereas endorphins manufactured by pancreatic cells do appear to reach the brain when given by injection.

Again, it must be stressed that this kind of research is still in its very early stages and a medically approved product is probably years away. In particular, it is very possible that endorphins and enkephalins will prove addictive, just as morphine is. This problem, and the need to target the drugs, might be tackled by chemically modifying them.

### Steroid hormones – an early success for biotechnology

While genetic engineering is the driving force behind the production of polypeptide hormones, biotechnology has already made a substantial contribution to the production of the quite different steroid hormones. Twenty years before genetic engineering became possible, a mould had brought relief to millions of people through the manufacture of cortisone.

The official birthplace of cortisone was an unusually august spot for a chemical compound – the ballroom of New York's Waldorf Astoria Hotel. In April 1949, the Mayo Clinic announced that a new medication for rheumatoid arthritis had been discovered, and it was known by the humble name of compound E. Very soon vitamin E began to be peddled as a cure for arthritis, despite the fact that its most notable connection with compound E was the presence of the same capital letter! In July, compound E was renamed cortisone and its illustrious career was launched.

The demand was immense, but supplies were extremely limited, and chemists set to work to find a way of making cortisone in bulk. They succeeded, but the process was complicated, entailing thirty-seven separate chemical reactions, and the cost of cortisone was accordingly high, $200 per g. Help, as so often, came from a microbe. In 1952, it was found that the bread mould, *Rhizopus arrhizus*, could convert another steroid, progesterone, into a compound from which cortisone could be made much more easily. The price of cortisone dropped to $6 per g, and by 1980, further improvements had reduced the cost to $0.46 per g. This was possible not only because the microbes circumvented a number of steps in the synthesis of cortisone (only eleven are now required) but also because the high temperatures and pressures, and expensive chemical solvents required for the chemical method of production could be avoided.

The case of cortisone illustrates very clearly that biotechnology and the methods of the chemical industry should not be viewed as all-or-nothing alternatives. The combination of chemical and biological processes is likely to play an ever-more important part in industry over the next decade.

The basic raw materials for the manufacture of cortisone and other steroid hormones are sterols, a group of compounds which are usually obtained from soya beans and the Mexican barbasco plant. As part of the process of making oestradiol and testosterone (both used in contraceptives) and other medically important steroids, two or more microbes are drafted in to make precise modifications to chemicals which are eventually converted into the final products.

Used with caution, steroids help alleviate inflammations, skin diseases, allergies and many other illnesses. Without microbes, many of these drugs would be totally unavailable or exceedingly expensive.

### Enzyme-replacement therapy

Each of us carries about half a dozen defective genes. We remain blissfully unaware of this startling fact unless we, or one of our close relatives, are among the millions of people who are afflicted by a genetic disease. Roughly one in ten individuals has, or will later develop, some inherited disorder, and approximately 2800 specific conditions are now known to be caused by defects (mutations) in just one of the patient's genes. Some single-gene diseases are fairly common – cystic fibrosis is found in one out of every 2500 babies born in Britain – and, in total, diseases which can be traced to single genetic defects account for about 5 per cent of all admissions to children's hospitals.

The reason why most of us do not suffer harmful effects from our several faulty genes is that each person has two copies of nearly all their genes, one derived from their father and one from their mother. The only exceptions to this rule are the genes found on the sex chromosomes in males. Males have one X and one Y chromosome, the former from the mother and the latter from the father, so each cell has only one copy of the genes on these chromosomes. In many cases, one 'good'* gene is

* One must be careful about using the terms 'good' and 'bad' in relation to genes. There is a considerable variation in the precise structure of human genes. 'Unusual' should certainly not be equated with 'bad'. Furthermore, all genes operate within a certain environment. Even genes which can produce definite diseases can have other beneficial effects. The classic example is the relationship between sickle-cell anaemia and malaria. Only individuals with two copies of the

sufficient to avoid all symptoms of disease. If the potentially harmful gene is recessive then its normal counterpart will carry out alone the tasks assigned to both. Only if we inherit from our parents two defective copies of the *same* recessive gene will a disease develop. Since our handful of gene defects are spread through thousands of different genes it is rather unlikely that you and your partner will carry the same faulty genes. So long as this unfortunate combination is avoided, your children will not suffer from recessive genetic diseases.

If the defective gene is dominant, it alone can produce the disease even if its counterpart is normal. Obviously only children of a parent with the disease can be affected, and even then on average half of the children will be unaffected. Huntington's chorea, a severe disease of the nervous system which only becomes apparent in adulthood, is an example of a dominant genetic disease.

Finally, there are X-linked genetic diseases, in which the defective gene lies on the X chromosome. As males have only one copy of the genes on this chromosome no others can step in to fulfil the defective gene's function. Both Duchenne's muscular dystrophy and haemophilia are X-linked.

Much research has focused on the so-called genetic metabolic diseases, in which a defective gene causes a particular enzyme to be either absent or ineffective, thus upsetting the delicate balance of the body's chemistry. Biotechnology can sometimes provide the missing enzymes.

One of the best-known inherited diseases is haemophilia, a disorder in which an individual's blood fails to clot properly. Queen Victoria was the most famous carrier of the defective gene responsible for this disease, and through her it was transmitted to the royal families of Russia, Spain and Prussia. Minor cuts and bruises, which would do little harm to most people, can prove fatal to haemophiliacs because they lack substances which promote the chemical reactions by which molecules in the blood link together to prevent further bleeding. Many haemophiliacs are now treated with proteins known as Factor VIII and Factor IX which are extracted from the blood of donors, and many biotechnologists are now trying to obtain these materials from genetically engineered microbes.

Gaucher's disease is one of the ten known hereditary lipid storage diseases, most of which cause extensive damage to the nervous system. Children suffering from this disease lack the enzyme which normally

---

sickle-cell gene, which produces defective blood proteins, suffer from this serious disease. Those with one sickle-cell gene and one copy of its normal counterpart are unaffected and, moreover, are able to resist infection by malarial parasites. The clear advantage of having one 'defective' gene explains why this gene is common among populations originating in areas where malaria is endemic. However, a move to a malaria-free area removes the 'advantage' of the 'defect', but its negative effects remain.

breaks down one of the many forms of lipids in the body, glucocere-broside. This material accumulates in the liver, spleen and bones, producing swellings, damage to nerves and, sometimes, death before the age of two. If the right enzyme could be delivered to the cells engorged with this lipid then the symptoms of Gaucher's disease could be relieved, and this is exactly what has been attempted. One of the scientists involved in this research, Roscoe Brady of the US National Institutes of Health, has said, 'I think our technology is on the verge of giving us the solutions we need. But until we can produce the required enzymes in much greater quantities and target them to specific areas, replacement therapy will have to be considered experimental'. Biotechnologists have developed a number of techniques for producing quantities of biological materials such as enzymes and targeting them.

At present the enzyme needed to combat Gaucher's disease is purified from human urine or placenta where it is present in very small quantities. It is time-consuming and expensive to collect sufficient enzyme even for clinical trials, let alone the full-scale therapy of many patients. Genetic engineers could solve this problem by persuading bacteria to manufacture it.

The specific biochemical problem, usually the lack of an enzyme or other protein, has been identified for about 200 of the known metabolic diseases. It is not yet possible to say how many can be treated by enzyme-replacement therapy, but during the next decade the picture should become much clearer. However, it is virtually certain that without the aid of biotechnology, doctors would have little hope of obtaining pure enzymes in large quantities at a reasonable cost – two of the prerequisites of any successful therapy of this type. We can be confident that as soon as the effectiveness of a particular enzyme-replacement therapy is demonstrated, biotechnologists will be able to meet the demand for the desired enzyme.

**Gene-replacement therapy – getting to the roots of inherited diseases**
Even the ambitious plans for enzyme-replacement therapy pale in comparison with gene replacement, one of the most radical notions ever put forward in medicine. Since inherited metabolic diseases result from absent or faulty genes, the ultimate cure would be to provide the patient with the correct genes which will enable the body to make the required enzymes or other proteins and so eliminate the root cause of the disease.

Despite the wide publicity gene-replacement therapy has received, the day when it becomes a practical proposition is far away. Although genetic engineers can merrily stitch all manner of genes into bacteria, gene-replacement therapy – like medicine in general – aims to affect something vastly more complex and subtle than mere cells; it is concerned with healing a whole individual consisting of billions of semi-independent cells.

Each one of the cells in the body carries the same genes, but different cells make different uses of the enormous variety of genetic information available to them. For example, skin cells have insulin genes, but do not use them; that task is left to specialized cells in the pancreas. Not merely do cells from different organs take on specialized roles, but the role of a specific cell can change with time. Most notably, the many changes that take place at puberty result from a whole panoply of genes being switched on or off in the cells of several organs.

The major practical obstacle barring the way to effective gene therapy is that there is, as yet, no sure method of introducing genes into human cells in such a way that they are subjected to the body's normal control systems. Clearly, if the gene remains switched off nothing has been accomplished. Equally, if it becomes too active – instructing the cell to make excessive amounts of a particular protein – the consequences could be as damaging as those of the disease it was intended to cure.

The only attempts at gene replacement in humans were made in 1980, when Martin Cline from the University of Los Angeles went to Italy and Israel to attempt to cure two patients suffering from thalassaemia, a severe and usually fatal blood disease. There are several different types of thalassaemia, all caused by defects in the gene coding for globin, a vital part of the haemoglobin molecule which gives blood its red colour and carries oxygen around the body.

Red blood cells, which are rich in haemoglobin, are derived from bone marrow cells. Accordingly, Cline and his colleagues removed some marrow cells from their patients and tried to introduce a normal globin gene into them (using vectors analogous to the plasmids employed in the genetic engineering of microbes) in the hope that this would take over the function of their own defective genes, allowing the patients to make normal haemoglobin. As globin genes were the first human genes to be cloned, there was no difficulty in obtaining normal genes for insertion into the marrow cells. After these cells had been genetically manipulated, they were replaced into the patients' bones.

These operations, however, failed, possibly because the genes entered the wrong part of the patients' chromosomes and did not begin making mRNA, the first step in the synthesis of a protein. Indeed, the endeavour provoked fierce controversy, since most experts considered the attempt at gene replacement to be very premature. These cases do, however, illustrate the principles behind gene replacement, which may eventually find a place in medicine.

Gene replacement is an ideal subject for armchair speculation. Is it possible that human beings could be designed to precise specifications? In most important respects the answer is, almost certainly, no, and for this we should doubtless be grateful. While it is quite conceivable that single genetic defects might be remedied, it is extremely unlikely that more substantial changes could be wrought on an individual's genetic make-up. It might be possible to correct a few of nature's errors, but there is little point in thinking we can improve on her best efforts, which are exemplified by the fortunate majority of us who do not suffer from a serious inherited disease. The manipulation of 'genes for intelligence' is pure fantasy. There is not even good evidence that what we choose to call intelligence (as measured through IQ tests) depends primarily on inherited factors. Even that part of intelligence which is genetically determined will almost certainly be a result of complex interactions between a multitude of discrete genes. The task of understanding such an intricate system is fearsomely difficult; the prospects for being able to tinker with it in a specific way are vanishingly small.

If gene-replacement therapy does become practicable, its first impact will be on the treatment of diseases in which the normal gene needs to be introduced into only one organ – that is, in cases where even in healthy people that gene is only expressed in cells of a particular tissue. Two such diseases are phenylketonuria (PKU) and sickle-cell anaemia.

PKU affects around one in 12,000 white children, and if untreated they become severely mentally retarded. The disease is caused by a defect in a gene which normally produces an enzyme in liver cells. PKU can be diagnosed soon after birth, and its worst effects are avoided if the children are given a special, and highly unpalatable, diet for their first few years.

Sickle-cell anaemia is widespread, predominantly among the black populations of tropical Africa and the US, and also in parts of the Middle East and Mediterranean areas. The disease is named after the characteristic shape of the patient's red blood cells. Like the thalassaemias, sickle-cell anaemia is caused by a fault in globin genes. Here again the hope is that bone marrow cells can be genetically engineered to enable the patients to manufacture normal globins, thus eliminating the excruciating pain and premature death associated with the disease.

In the next decade or so it should become clearer whether gene-replacement therapy will have a significant role to play in twenty-first-century medicine. Until then we must continue to rely on conventional treatments and genetic counselling, in which couples who both carry the same deleterious genes are advised on their chances of producing affected children.

## Scourges of the affluent

While most of the diseases discussed so far in this chapter tend to afflict all age groups, but often the young in particular, cancer and cardiovascular diseases primarily strike mature adults and the elderly. Cardiovascular diseases, particularly heart attacks (caused by a blockage of the blood vessels that supply oxygen to the heart), and strokes (blockage of blood vessels around the brain), account for more than half of all deaths in developed countries, such as the US and UK. The National Science Foundation estimates that about one person in four now living in the US will develop some form of cancer and, on current trends, one in six will die from it.

The scale of the problem needs no emphasizing and an attack on these two groups of diseases, particularly cancer, is a prime aim of many biotechnologists. The challenge is formidable, not least because the causes of these diseases are far from clear. There are, however, some encouraging signs that new methods of diagnosis and treatment, based on biotechnology, will have an impact before too long.

### Cancer and the roots of malignancy
Among the relatively affluent third of the world's population no disease evokes as much dread as cancer. The insidious and painful nature of most cancers amply justifies the public's concern. This disquiet has prompted many a drive to 'find a cure for cancer', most notably the US National Cancer Act of 1971 which pledged no less than $600 million to the crusade against cancer. Why, then, in the face of a massive and concerted attack from many of the world's finest scientists and doctors, has cancer remained so intractable?

In fact, cancer research has been more successful than is often recognized. For example, early diagnosis and improved therapy have made the outlook much brighter in cases of cancer of the cervix and some leukemias. However, the death rates for lung cancer, the most prevalent type of cancer in British and American males, and breast cancer, the most frequent in females, have remained obstinately high.

This slow progress has led to strident enquiries about why there is not a cure for cancer around the corner, similar to that for polio. This brings us to the nub of the problem – cancer is not a single disease. Polio is caused by identifiable viruses and this made possible the production of highly effective vaccines to combat the disease. The chances of finding an analogous anti-cancer vaccine are virtually nil. Cancer has many causes and takes many forms.

When our bodies are functioning properly, every cell and organ forms part of a precisely coordinated and regulated system. The whole process of growth, from a single fertilized egg cell to a mature adult, depends on cells growing, dividing and taking on specialized tasks according to an intricate pattern. Once an organ or other group of specialized cells has reached its correct size, signals are generated in a way which is not understood in detail and cell growth largely ceases. From then on new cells are normally generated at a slow rate, just sufficient to replace those that are dying. In cancerous cells, the usual regulation has gone haywire; they continue to grow and divide, invading areas of healthy tissue with disastrous results.

The obvious problem which faces cancer researchers is to determine how cancer cells escape the body's normally stringent control mechanisms. Many details have been uncovered about the molecular processes that seem to be deranged in such cells, but a coherent and comprehensive picture has yet to emerge. Rather more extensive knowledge exists about the agents responsible for transforming cells.

About 500 different chemicals are known to be carcinogenic (cancer-producing) in animals, a few dozen of which have been implicated in human cancers. Obviously the search for human carcinogens is greatly impeded by the ethical impossibility of testing chemicals to see if they cause cancer in humans. Thus, most of the firm evidence about human carcinogens is gathered by looking at the incidence of specific types of cancer in people whose work or habits expose them to particular chemicals. A strong link between vinyl chloride, a chemical used to make the plastic PVC, and liver cancer has been established, and the link between cigarette smoking and lung cancer seems incontrovertible, to mention just two of the best known examples. Materials as diverse as benzene, asbestos and chromium have also been implicated in human cancer. A fierce debate rages around the importance of these environmental factors in cancers, with some experts claiming that 85 per cent of all cancers can be traced to them. In any event, 'industrial' carcinogens are unlikely to account for many cancers in the population as a whole, since most people do not come into contact with them.

It is a daunting task to try to unravel the plethora of claims relating life-style to cancer. Certainly, particular dietary habits appear to be associated with some types of cancer – for example, lack of fibre with cancer of the colon, and the high incidence of stomach cancer prevalent among those on traditional Japanese diets. It is questionable, however, that such findings can form the basis for a 'cancer-preventive diet'. In its 5-Year Outlook on Science and Technology, published in 1982, the US National Science Foundation notes that the 'avoidance of one form of risk seems almost necessarily to engender another'.

Some radiations, including those produced by nuclear reactions and ultraviolet radiation in sunlight, clearly cause cancer. Sunlight is the major cause of skin cancer, but fortunately this is one of the most effectively treated forms. There is, however, a deep controversy about the amount of radiation that we can tolerate without harm. Everyone agrees that massive doses of radiation induce cancers, but opinions differ on the danger of the much smaller amounts that the general population is likely to encounter. Some experts assert that the incidence of cancer is directly proportional to the amount of radiation and that even small doses will cause some cancers; others aver that there is a threshold level, below which no harm is done. It may take many years before one side or the other has assembled enough evidence to establish its case. In the meantime, prudence seems to favour the assumption that all levels of radiation are potentially damaging.

The final important suspects in the production of cancer are viruses. Nearly every major university is engaged in the study of viruses that cause tumours in animals. Their combined efforts have revealed intricate details about the way viruses subvert a healthy cell. Dozens of specific viruses are known to transform normal cells into cancers in mice, rabbits, chickens, monkeys and other animals. However, only four forms of human cancer have been strongly linked with viruses – a type of liver cancer with hepatitis viruses, two rare forms of blood cell cancer, and a cancer of the nose and throat found mainly in Africa. A few years ago it seemed that viruses might be the key to cancer in general, perhaps even that there was a 'human-cancer virus'. That idea fell into disfavour, but with the very recent research on oncogenes it is beginning to reappear in a new guise.

It seems that oncogenes are sections of DNA present in every cell – healthy as well as cancerous. They seem to lie dormant until activated, when they transform the cell into a cancerous form, initiating the headlong proliferation typical of the disease. Intriguingly, their structure is very similar to that of certain viruses, so perhaps there is a relatively small number of human-cancer 'viruses', although these 'viruses' are an intrinsic part of our genetic inheritance and not extraneous infections. If this theory stands up under closer scrutiny, there may at last be a unifying theme in cancer research, one which has been sorely lacking until now. The task will then be to discover just how these oncogenes are roused from their dormant state to exert their malignant influence. This may result in the apparently disparate 'causes' of cancer outlined above being linked by their shared ability to activate oncogenes.

These ideas have profound implications for cancer research. Paradoxically, they would mean that despite the crucial role of virus-like material in the induction of cancer, there would be no hope of manufacturing an all-purpose, anti-viral vaccine to protect against the disease. Vaccines can only be used to boost the body's defences against *foreign* materials, not against indigenous components of cells. On a more optimistic note, it is conceivable that ways could be developed of permanently inactivating oncogenes, thus hitting at the very roots of cancer. At present this is certainly science fiction, not science fact, but the pace of modern molecular biology is so rapid that one can rarely say that fiction will never turn into fact.

This catalogue of largely unanswered questions about the nature and causes of cancer may seem a gloomy background against which to set the current endeavours of biotechnologists. However, substantial advances can be made, even in a field so beset with theoretical and practical problems. One indication of such progress came in the summer of 1983 when scientists from Britain and the US announced that they had discovered that one of the known oncogenes contains the instructions for manufacturing a substance called platelet-derived growth factor (PDGF).

This substance is normally synthesized by blood cells in response to wounds and it prompts the growth of new cells to repair the damage. Clearly, if growth-stimulating substances such as PDGF were manufactured by other types of cells they might be induced to grow uncontrollably. The discovery of the gene for PDGF in oncogenes may provide a clue to the development of cancers.

### Interferon – hopes and hypes, facts and fantasies

Few drugs have received a barrage of publicity comparable to that showered on interferon. Stories of precious vials of interferon being flown around the world to offer hope to cancer sufferers have bolstered a widespread belief that interferon can knock out cancer.

In the face of the media hype – sometimes supported by reckless quotes from scientists and doctors – it is tempting to over-react and dismiss interferon as a genuine aid in cancer therapy. The bald fact is that no-one has proved unequivocally that interferon can cure any form of cancer. Until very recently there was not enough interferon available in the world for proper clinical trials to test its efficacy. This situation is rapidly changing. Genetic engineers are supplying more and more interferon, and a number of level-headed institutions are funding research into its effects; for example, the American Cancer Society is spending $5 million and Britain's Imperial Cancer Research Fund has allocated about £1 million.

Encouraging signs are now emerging. Interferon seems to benefit patients suffering from cancers of the skin, bone, breast and blood. The next few years will see a steady stream of scientific publications presenting the results of tests with several different types of interferon in patients with many forms of cancer. Only when this detailed information has been accumulated will it be possible to say what role, if any, interferon will have in cancer therapy in the future. At present, the best guess is that it will have some part to play, not as a miracle cure, but as an extra weapon in the clinician's armoury.

One new line of research concerns the notion of producing hybrid interferons. The dozen or more types of interferon fall into three classes – alpha, beta and gamma – according to the type of cell that produces them in greatest abundance. Alpha and beta interferons are chemically similar and are derived from leucocytes (a type of white blood cell) and fibroblast cells from connective tissue of muscle and skin. Gamma interferon is rather different and is obtained from cells of the immune system. By carefully snipping the genes that code for interferons it will be possible to induce bacteria to manufacture a hybrid interferon consisting, for example, of the first half of one form joined to the second half of another. It could then be tested to see if it is more effective than either of the natural products.

One of the many surprises thrown up by research into interferon concerns the role of sugar molecules which are attached to most interferons manufactured in human cells. Like many proteins, interferons are not simple chains of amino acids, but also have sugar groups tacked on to the chain at certain points. Proteins like these are called glycoproteins. It would be difficult indeed for bacteria to make them – they would not 'know' where or how to affix the sugar groups and genetic engineers are a long way away from working out how to tell them. Fortunately, interferons which lack the sugars and can be made by suitably instructed microbes seem to act in just the same way as their natural glycoprotein counterparts.

The relatively scanty understanding of interferons appears rich indeed compared with knowledge about two other classes of molecules that are attracting increasing attention from cancer researchers. Lymphokines, a group of about 100 substances, are produced by white blood cells while some cytokines are associated with the thymus gland. The US National Cancer Institute has concluded that several of these materials may have great potential in cancer treatment. Since many of them are glycoproteins, and for these the sugar groups may be essential, there is clearly an incentive to find out how to add sugar groups to genetically engineered proteins. Chemists and biochemists, rather than genetic engineers, are most likely to come up with the answers.

## Monoclonal antibodies home in on cancers

Before cancer cells can be attacked with antibodies it is essential to find out what distinguishes them from normal cells. Only then can antibodies be manufactured which will home in on malignant cells and ignore all others. This is a formidable task, but substantial progress has already been made.

The root of the problem is simple to understand – cancer cells *are* human cells, albeit ones which have taken a treacherous turn. It is not surprising, therefore, that the surfaces of cancer cells (where the antigens lie) are very similar to healthy cells. Fortunately, the inner turmoil of cancer cells is reflected in subtle changes to their surface antigens. This has meant that, in recent years, monoclonal antibodies which pick out these characteristic antigens have been developed. No one has yet managed to find an antigen which is completely cancer-specific – that is, one which is found on cancer cells and absolutely nowhere else. However, in the summer of 1982 exciting news came out of Oxford University. Henry Harris and his colleagues published their investigations into the Ca antigen, a feature which seems to be found only rarely on non-cancerous cells. If this proves to be the case then monoclonal antibodies against the Ca antigen could prove invaluable in the diagnosis and treatment of cancer.

It is now possible to make monoclonal antibodies which recognize antigens characteristic of (but not entirely specific for) several types of cancer, including three of the major killers, cancer of the breast, lung and bowel. This list, which also includes leukemias and cancers of the skin and pancreas, will surely be extended very rapidly. The immense labour involved can be judged from the fact that researchers at the US National Cancer Institute had to examine over 15,000 different types of monoclonal antibodies to find one which is specific for a form of lung cancer that strikes about 32,000 Americans each year.

At present, monoclonal antibodies are used almost entirely for diagnosis and monitoring the effectiveness of conventional forms of cancer therapy. The success of a treatment can be measured by the decrease in the number of cancer markers detected by the appropriate antibodies in a sample of the patient's blood. The therapeutic applications of antibodies are still in their infancy. The most encouraging news so far has come from Ronald Levy and his colleagues at Stanford University School of Medicine. Their preliminary reports suggest that a cancer of, as it happens, antibody-producing cells (B-cell lymphoma) may be treated with monoclonal antibodies.

Monoclonal antibodies may, by themselves, trigger the patient's own immune system to start attacking a cancer, but it is a refinement of this idea which is creating interest among cancer specialists. This is to use antibodies as 'guided missiles' to deliver a toxic 'warhead' straight on to cancer cells. The majority of drugs distribute themselves fairly evenly throughout the patient's body, but in many cases they are really only needed at one or a few specific sites, namely where the cancer has taken hold. Many of the most powerful drugs produce harmful side-effects in the patient by affecting normal cells as well as diseased ones, and this is particularly true for anti-cancer drugs. It would clearly be desirable to give them in smaller doses, while making sure that they reach the spots where they can do most good. This is just the facility offered by monoclonal antibodies. It seems entirely possible that monoclonal antibodies targeted against antigens found only on cancer cells can be linked to anti-cancer drugs, delivering them directly to the cancer. The beauty of this technique is further enhanced by the fact that it would not even be necessary for doctors to know *where* all the cancers are – the monoclonals would find them.

Caird Edwards and Philip Thorpe of London's Chester Beatty Research Institute have called this a retiarian approach to cancer therapy. The retiarii of ancient Rome were gladiators armed only with a net and trident. Monoclonal antibodies carrying anti-cancer drugs could work in a similar manner – the antibodies would ensnare cancer cells and then the toxic drugs would deliver the *coup de grâce*.

The possibility of targeting drugs very precisely would open the way to a whole new range of anti-cancer agents. Since the anti-cancer drugs cannot be sent directly towards cancers, only relatively non-toxic substances can be employed. (The word 'relatively' is significant here. All effective anti-cancer drugs cause considerable damage to other parts of the body.) If a drug linked to a monoclonal antibody could zero in on cancer cells only, then far more lethal cell poisons could be used. One such is ricin, a protein from the humble castor bean plant. Ricin is so potent that one molecule is enough to kill a whole cell.* Recent experiments in Dallas, Texas, showed that ricin linked to antibodies killed cancer cells in the bone marrow of mice without destroying healthy cells.

In fact, in application ricin can be made less dangerous. Ricin, like many other toxins such as diphtheria and cholera toxins, consists of two amino acid chains linked by a single bridge. One, the B chain, helps the ricin molecule cross the membranes that surround cells, while the other, the A chain, is the part which does the killing. By using only the toxic A chain joined to an antibody, the risk that ricin will get into healthy cells is reduced (see Figure 6).

Obviously the effectiveness of ricin and similar poisons depends on their entering the cell under attack. With the removal of the B chain how will ricin A chain get in to carry out its destructive role? The complete answer is not known, but part of it is an exquisite example of nature's cunning – and how it can sometimes backfire. When an ordinary antibody latches on to an antigen on a cell's surface, the antibody causes no direct harm to the cell, but it does act as a signal to other parts of the immune system which destroy the cell. One way the 'tagged' cell can avoid this fate is to engulf the antibody so that it can no longer alert the immune system. This avoids trouble if an ordinary antibody is involved. However, if the antibody carries with it a toxin such as ricin, the cell, like the Trojans who welcomed the Greek horse, has ensured its own destruction.

At present nearly all the monoclonal antibodies used in research are derived from mouse cells and a few from rats. Incredibly useful though these are, they do suffer from one important disadvantage: because they come from a different organism, the immune system of a human patient may produce its own antibodies to attack the mouse monoclonals. To prevent this futile battle between antibodies, it is necessary to make human monoclonal antibodies in the laboratory. At present this cannot be done efficiently, but there is every reason to believe that it may soon be achieved. The main requirement is for a suitable human myeloma cancer to be found, one which is analogous to the mouse myeloma cells already available. This could then be fused with spleen cells to provide hybridomas manufacturing human monoclonal antibodies.

In the meantime, the use of mouse antibodies in diagnosis presents no danger, because the antibodies do not need to be introduced into a

* Ricin's first taste of fame, or rather notoriety, was in the bizarre murder case of Georgi Markov, the Bulgarian broadcaster working in London. After his death a tiny metal capsule was discovered in his leg. This capsule had probably been filled with a droplet of ricin and fired into him by an air-gun disguised as an umbrella.

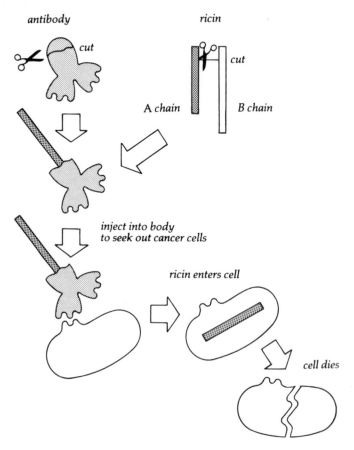

**Figure 6**
Using monoclonal antibodies to guide poisons to cancer cells. The poison ricin is composed of two amino acid chains. The A chain, which is the part responsible for its toxic effects, is attached to a monoclonal antibody that recognizes antigens on cancer cells. When the antibody/A-chain combination is injected into the body it will seek out cancer cells and kill them.

patient, but simply mixed with samples of blood or other substances from the patient. For severely ill patients, it may be worth taking the risk of injecting mouse antibodies if they offer some significant chance of a cure.

**Enzymes undermine cancers**
Monoclonal antibodies rely on differences in the *structure* of cancer cells.

Another emerging biotechnology depends on differences in the way cancers actually *work* – that is the chemical reactions which go on inside the cell, or its metabolism. Again, the great similarity between cancerous and normal cells must be stressed, but a few potentially important differences have been unearthed. One serves to illustrate another way that biotechnologists can help combat cancer.

Asparagine is one of the twenty amino acids used to construct proteins, and without it no cell can live for long. Normal human cells can manufacture their own asparagine from simple and readily available chemicals in their environments. Certain types of cancer cell, however, lack this ability, and they must pick up the asparagine they need from the bloodstream. Any discrepancy of this kind immediately excites the attention of cancer researchers, and this prompted ideas of cutting off the vital supply of asparagine to cancer cells and thus starving them to death.

There is an enzyme, known as asparaginase, which selectively destroys asparagine molecules. Asparaginase is found in many diverse organisms, including yeasts, bacteria and plants. Most of the medical supplies are now obtained from four types of bacteria, the most important of which is *Erwinia*. Purified asparaginase has been injected into children suffering from the blood cancer acute lymphoblastic leukemia, and about half of the children improved significantly after this treatment. Presumably the asparaginase had cleared most or all of the asparagine from their blood, with disastrous effects on cancer cells but little on normal ones. Asparaginase might also be effective against other types of cancer, but this has not yet been proved.

Obviously the biotechnologist's task is to design efficient fermentations to provide clinicians with as much asparaginase as possible at the lowest cost. In particular, the yield of asparaginase from the bacteria growing in large vats is increased by adding various simple chemicals. The most effective seem to be two other amino acids, leucine and methionine. It is not really known why these amino acids cause the bacteria to make more asparaginase. This kind of problem in biotechnology is still usually tackled by trial and error.

Clearly this approach towards cancer therapy could take many forms. The search for other 'metabolic markers' of cancer cells continues. Once discovered, drugs can be designed which exploit characteristic weaknesses of these otherwise all too vigorous cells.

Quite a different use of microbial enzymes is being developed to aid doctors in the treatment of cancer. Methotrexate (MTX) is a widely used anti-cancer drug which works by damaging an enzyme called dihydrofolate reductase. This enzyme is involved in the manufacture of several compounds that cancer cells, in particular, need to survive. However, MTX also damages healthy cells so it is vital that only the minimum effective dose is given to patients. Since different patients break down the drug at different rates it is important to know just how much MTX is in their bodies at any time. The amount of MTX in the blood can be measured by testing its effects on dihydrofolate reductase, and the enzyme used in this test is obtained comparatively cheaply from a bacterium called *Lactobacillus casei*.

Undoubtedly microbial enzymes will continue to grow in importance in medicine as the inherent advantages of microbes as excellent sources of many diverse materials become fully exploited. According to one estimate, microbial enzymes will be worth 60 per cent of the total $1 billion diagnostic enzyme market that will exist by the end of the eighties.

## Organ transplants

No recent medical advance has received more public attention than organ transplants, especially of hearts and kidneys. Admiration for the delicate skills of surgeons has rather overshadowed the crucial contributions of immunologists. Somehow the patient's body must be persuaded to accept a foreign object – the transplanted organ – and not assault it with the immune system. A number of drugs which suppress the body's immune system have been developed in recent years and one of the best is Cyclosporin A, a compound made by the fungus *Tolypocladium inflatum*. This organism, which was first discovered in a sample of Norwegian soil, is set to revolutionize organ transplantation. Without Cyclosporin A only about 35 per cent of patients who receive liver transplants survive for a year; with the drug the figure nearly doubles. Similarly, trials of Cyclosporin A have increased the success of kidney transplants from 50 per cent to over 80 per cent.

Cyclosporin A works by affecting the 'T cells' of the immune system, and its uses appear to extend far beyond organ transplantation. There are many diseases which seem to be caused when the body is attacked by its own immune system. One such 'auto-immune disease', uveitis, an inflammation of the eye which can result in blindness, has been treated successfully with Cyclosporin. The drug has also shown considerable promise in killing schistosomiasis and malarial parasites (see p. 88). There are even indications that it can help control some types of diabetes. When one considers that Cyclosporin A has only been available to a few scientists and doctors for five years or so, it is remarkable how many beneficial effects it has revealed. Cyclosporin A could clearly become one of biotechnology's greatest gifts.

Drugs which are given to suppress the immune system and so help prevent organ rejection also increase the risk that the transplant patients will catch infectious diseases, in particular, kidney transplant patients sometimes suffer from herpes viruses. Interferon could play a vital role in guarding against this.

Monoclonal antibodies could also help to supplement the patient's debilitated defences, but more importantly monoclonals can be used to select the best possible organ – that is, one which is as similar as possible –

for transplanting into a particular patient. Tissue typing, as the science of matching organs is known, will be revolutionized by monoclonal antibodies. These will identify the antigens on potential donor organs and allow doctors to select those that are most similar to the patient's own organs. This approach goes right to the root of the problem of rejection, since it is foreign antigens which prompt the patient's immune system to fight against the new organ. Ensuring that transplants have as few foreign antigens as possible should reduce the need for drugs and, thus, keep the patient's immune system in better shape.

## Cardiovascular disease

As we have all observed, blood around a wound quickly solidifies into a clot which prevents further blood loss. Clearly this is a vital process; if this clotting ability were absent or reduced, as it is in haemophiliacs, a minor cut could prove fatal. Equally, however, it is essential that clots do not form inside the veins and arteries where they would halt the free flow of blood carrying oxygen to the tissues. If clots do form the effects can be disastrous – a heart attack if the obstruction is near the heart, and a stroke if it is near the brain.

Many substances are involved in the body's control of the clotting process. Some substances promote clotting, while others retard it. This system of checks and balances is a marvellous example of biological regulation, but sometimes, for many different reasons, blood vessels do become blocked, and two products of biotechnology can help unplug them, minimizing the damage done.

In the nineteen-thirties, intense research began into an enzyme produced by *Streptococcus* bacteria. This enzyme, streptokinase, has been used since the seventies to clear obstructions in leg and lung veins, and, in 1982, the US Food and Drug Administration approved it for treating heart attacks. Promising though streptokinase is, it suffers from a drawback common to the great majority of bacterial enzymes that are introduced into a patient's body – it can provoke an immune response which diminishes its effects. Attention is, therefore, shifting towards an enzyme which has similar properties but is made by human cells – urokinase.

Much of the world's supply of urokinase comes from Japan, where annual sales have already reached £100 million. At present, urokinase is either extracted from urine or obtained from kidney cells in laboratory cultures. Both processes are exceedingly expensive and this fact, together with the growing interest in using it to treat blood clots in Japan and Europe, has motivated genetic engineers to try to clone the relevant gene and introduce it into bacteria. At least two groups have claimed success, but their subsequent progress in developing a commercial fermentation system for producing the drug is shrouded in mystery. In theory at least there is no reason why genetically engineered urokinase and other enzymes involved in the breakdown of blood clots should not be generally available by the mid-to-late eighties.

A group of compounds known as tissue plasminogen-activators (tPAs) have attracted even greater attention in the last two years. Urokinase seems to act by decreasing the clotting ability of blood throughout the body, with the consequent risk of internal bleeding. The tPAs, by contrast, appear to be more precise in their action, attaching themselves to blood clots and stimulating other components of the blood to break down the clot, without reducing the blood's clotting power elsewhere in the body. Some tPAs have now been cloned and will soon be tested.

An unlikely alliance of fireflies and bacteria is beginning to find a place in the diagnosis of heart and liver disease. The wakes created by ships as they plough through the oceans, especially in tropical waters, are frequently marked by a strange shimmering light. This bioluminescence comes from millions of luminous bacteria in the water. Both these bacteria and fireflies produce light by chemical reactions involving enzymes.

Increasing amounts of fatty substances, known as triglycerides, in the bloodstream warn of the risk of atherosclerosis, the narrowing of arteries caused by a build-up of fatty materials on the inside of the vessel walls. Although there are already several methods of measuring the amount of triglycerides, a newly developed bioluminescent approach is more accurate and faster, and may be cheaper.

Luminescent bacteria give off light when the enzyme luciferase catalyses a particular type of chemical reaction. This reaction can only take place in the presence of an energy-rich compound, known by its initials as NAD(P)H. Thus, the amount of NAD(P)H can be determined by the light emitted from a solution which contains luciferase extracted from bacteria. When blood containing triglycerides is mixed with another type of enzyme and NAD(P)H, the concentration of NAD(P)H in the solution changes according to the quantity of triglycerides present. This changes the amount of light given off by the solution, giving an indirect but accurate measure of the triglycerides in the blood sample. Similarly, firefly luciferase can be used to show the concentration of creatine kinase (which increases after a heart attack) and other materials.

Similar methods have been developed for measuring alcohol, certain hormones and bile acids (which indicate liver damage). Furthermore this approach could be employed to analyse 200 or more substances in the body if these are found to be significant indicators of various diseases.

The prevention, diagnosis and treatment of cancers and diseases of the heart and blood vessels are certainly among the largest challenges facing biotechnologists today. Few major advances have yet been made, but encouraging progress in several areas provides the hope that the next decade will see significant improvements in our ability to combat these diseases, which now kill over half the population in the developed world.

# CHAPTER FIVE

# The new green revolution

Between 1930 and 1975, the average yield of corn per hectare in the US more than tripled. This remarkable increase in productivity is just one of the many achievements of what came to be known as the green revolution. The development of higher yielding varieties of corn, wheat, rice and many other crops, allied to the greater use of irrigation, fertilizers, pesticides and herbicides, have played a large part in ensuring that the world's food shortages are not even more acute. While the green revolution has not been an unmitigated success – for instance, the adverse environmental effects of pesticides cannot be denied – there is little doubt that it has made a major contribution to human welfare. However, the last few years have seen a considerable slowing down of yield increases as the green revolution runs out of steam. Biotechnologists, particularly genetic engineers, are hopeful that they can revitalize many sections of agriculture by enabling us to grow more food at a lower cost and, perhaps, to utilize land that lies idle at present.

## Nitrogen – a key to plant productivity

Whenever a complicated and coordinated structure is being built up, there is often one component that is limiting. For example, it is pointless in a motor-car factory to employ more people or increase the speed of the assembly line if there are only enough spark plugs to complete the present output of vehicles. In the case of crop plants, the limit to growth is often set by the availability of nitrogen in the soil. So, there is little value in increasing the irrigation in a field of wheat if the soil lacks enough nitrogen to support extra plant growth.

Nitrogen is required to construct most of the vital compounds inside cells, including proteins, DNA and RNA. Animals obtain the nitrogen they need by eating plants or other animals, whereas plants must extract it from the soil. A massive industry has grown up to meet the vast and rapidly increasing demand for fertilizers which supplement the land's natural supplies of nitrogen, but the costs are high and likely to increase substantially in the next few years. American farmers spend nearly $1 billion a year on nitrogen fertilizers for the corn crop alone, and the 40 million tonnes now spread over that nation's fields might need to quadruple by the end of the century if present trends continue. The price rises can only be guessed at, but it is highly significant that the fertilizer industry depends heavily on oil, consuming about 10 per cent of the US total. The urgent need for alternatives to nitrogenous fertilizers is obvious, and 'green gene' engineering could supply an answer.

Figure 1 shows the nitrogen cycle, the global processes by which nitrogen continuously moves between plants, animals, microbes, the land and the atmosphere. It may seem strange that there can be a shortage of nitrogen when 80 per cent of the air that surrounds us consists of nitrogen gas, but for most forms of life nitrogen gas ($N_2$) is totally useless. Plants, animals and most microbes have no way of incorporating gaseous nitrogen into their cellular compounds. Plants and most microbes generally obtain their nitrogen from compounds such as ammonia ($NH_3$) and nitrates ($NO_3$). The nitrogen in these and other substances is said to be in a fixed form, in contrast to the free nitrogen in $N_2$ gas. Many biotechnologists in universities and some of the largest corporations are focusing their attention on the relationship between plants and certain types of microbe which can fix nitrogen – that is, microbes that extract nitrogen gas from the air and incorporate the nitrogen atoms into molecules of ammonia, nitrates or other compounds which plants *can* utilize.

From the time of the Romans, and probably earlier, it has been known that planting legumes, such as peas, beans and clover, in fields which have previously been used for crops such as wheat restores the fields' fertility. The reason for this is now known: billions of tiny bacteria live on the roots of legumes and fix atmospheric nitrogen, so helping to replace the nitrogen extracted from the soil by the previous years' crops. There is no finer example of the subtle interactions between the millions of species that constitute the web of life on Earth than the relationship between legume plants and nitrogen-fixing bacteria. Both plant and microbe benefit from each other's presence; microbes obtain energy-rich nutrients from the plant and, in return, supply the nitrogen-containing compounds which the legumes need to maintain their growth. Many such symbiotic

---

**Figure 1**

The nitrogen cycle is the movement of nitrogen between plants, animals, microbes, the land and the atmosphere. Nitrogen exists either in the form of nitrogen gas ($N_2$) or fixed into a wide variety of chemical compounds, including proteins, ammonia and nitrate.

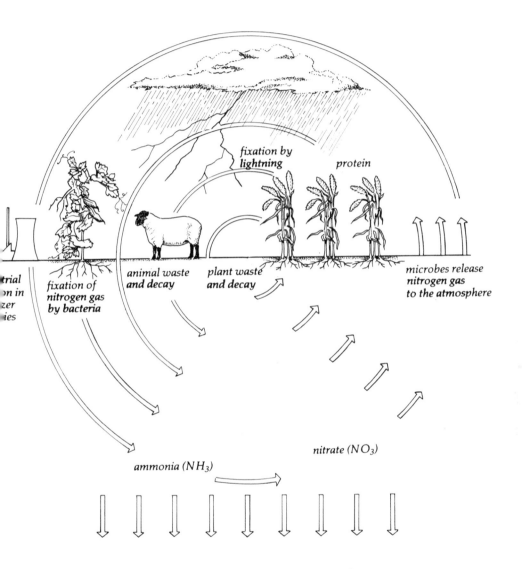

nitrogen gas in the air

fixation by
lightning

protein

trial
on in
zer
ies

fixation of
nitrogen gas
by bacteria

animal waste
and decay

plant waste
and decay

microbes release
nitrogen gas
to the atmosphere

nitrate ($NO_3$)

ammonia ($NH_3$)

leaching of nitrogen compounds into the earth

(mutually beneficial) relationships are known in nature. So far, we have had to rely on the legume/nitrogen-fixing bacteria association as it is presented to us by millions of years of evolution, but during the next decade or so it is quite possible that the phenomenon may be manipulated to bring even greater benefits to agriculture.

Some legumes are of great economic importance, for example, peanuts and soybeans. However, most of the world's major crops – rice, wheat, corn, barley and other cereals – must be grown without the direct assistance of nitrogen-fixing bacteria which do not find these crops' roots a congenial place to live. Biotechnologists can offer three feasible means of helping these crops reap the rewards offered by the microbial fertilizer factories. First, an attempt could be made to modify either the microbes or the wheat, or both, so that each could benefit from an association with the other. Second, it may be possible to modify other types of bacteria that already live contentedly on wheat so that they, too, could fix nitrogen. Finally, and most ambitiously, one could try to genetically engineer new types of wheat which could fix their own nitrogen from the air, by transferring genes from existing nitrogen-fixing microbes.

The nitrogen-fixing microbes found in the roots of legumes are various species of *Rhizobium* bacteria. These rod-shaped organisms infiltrate the root hairs and take up residence in the root itself, producing nodules which are characteristic of legumes. The subtle interactions between the bacteria and their plant hosts are exceedingly difficult to unravel. At present there is not a full understanding of the way that *Rhizobium* enter the root hairs, penetrate into the body of the root and form nodules, receive the plants' supply of nutrients and make their 'payment' of nitrogen-containing compounds. Before persuading *Rhizobium* to take up new residences, say in wheat roots, more needs to be learned about their natural living conditions so that these can be matched as closely as possible. This is a formidable challenge and, at the moment, the other two possible routes to the goal seem more straightforward.

In the last decade or so, an immense amount has been learned about the molecular machinery used by bacteria to fix nitrogen. The central feature is an enzyme called nitrogenase. This takes nitrogen gas and, using energy largely derived from the plant host's photosynthetic activities, converts the gas into ammonia. More than a dozen genes, termed the *nif* genes, are involved in the assembly of the nitrogen-fixation apparatus. At first sight, it would appear to be a herculean task to transfer all these genes into another type of microbe. Fortunately, as is often the case with genes that are involved in performing a single function within cells, these *nif* genes are linked – that is, they are not scattered throughout the vast amount of DNA that makes up the bacterial chromosome, but are all clustered together in one region. This makes it much easier to cut out the relevant stretch of DNA in a *Rhizobium* chromosome and insert the whole batch into another organism.

Genetic engineers have already managed to transfer the *nif* genes from a nitrogen-fixing bacterium into *E. coli* and, equally importantly, the *E. coli* was then able to fix nitrogen. This experiment did not use *Rhizobium* genes, but *nif* genes taken from *Klebsiella pneumoniae*, a soil bacterium which lives independently of any plant host. This bacterium has no less than seventeen *nif* genes, and the fact that it was possible to transfer all of these into a new home augurs well for future work on the bacteria which now colonize the roots of wheat and other cereals but cannot fix nitrogen.

More enticing still is the prospect of inserting *nif* genes directly into crop plants, dispensing with the need for any sort of nitrogen-fixing microbe. In this approach, a number of serious problems are encountered, not least in duping a plant cell into treating bacterial genes as its own. As mentioned earlier, the most significant division in all forms of life is between the eukaryotes (in which DNA is packaged inside a nucleus) and the prokaryotes (which have no nucleus). At the molecular level this distinction is far more important than the more obvious differences between two eukaryotic organisms, such as a rat and a tomato plant. Since all bacteria are prokaryotes and all plants are eukaryotes, the bacterial *nif* genes will not be 'understood' by the crop plants. In particular, ways must be found of ensuring that the plants manufacture the right amounts of the proteins specified by the bacterial *nif* genes.

This presupposes that genes can actually be transferred from bacteria to plants. Many examples of genetic engineering have already been described which involve the transfer of genes from eukaryotes to prokaryotes (for example, human insulin genes into *E. coli*) and from eukaryotes into other eukaryotes (for example, interferon genes into yeasts). These techniques are well developed in comparison with the transfer of prokaryotic genes into eukaryotes, for example, from *Rhizobium* into wheat. In the last few years, however, great advances have been made in the understanding and application of eukaryotic vectors – pieces of DNA that can ferry foreign DNA into eukaryotic cells. As with the vectors employed to carry genes into bacteria, there are two main types, viruses and plasmids, both of which are becoming increasingly employed in plant genetic engineering.

The key to many recent developments is to be found in a strange excrescence called a crown gall (see Plate 8), a type of tumour which afflicts many flowering plants. The lumpy tumour itself consists of a mass of plant cells which rapidly proliferate because they have escaped the plant's normal growth control mechanisms and, in this sense, they are analogous to animal tumours. However, in the case of crown galls the causative agent is a bacterium, *Agrobacterium tumefaciens*, inside which are small circular pieces of DNA known as Ti (tumour-inducing) plasmids. A crown gall is created when Ti plasmids are transferred from the bacteria into the chromosomes of the infected plant, the resultant change in the genetic constitution of the plant cells inducing them to grow and divide very

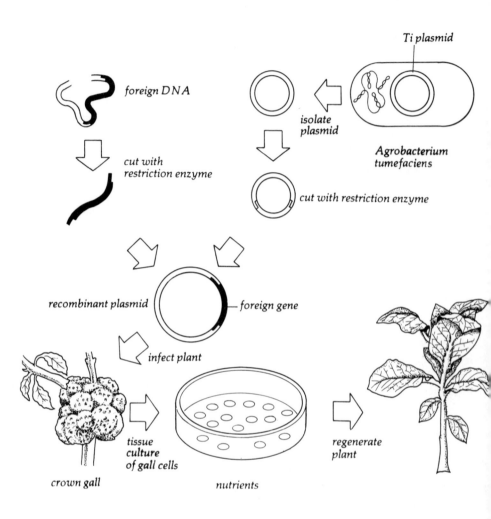

foreign DNA

Ti plasmid

isolate
plasmid

*Agrobacterium
tumefaciens*

cut with
restriction enzyme

cut with restriction enzyme

recombinant plasmid

foreign gene

infect plant

crown gall

tissue
culture
of gall cells

nutrients

regenerate
plant

**Figure 2**
Plant genetic engineering with Ti plasmids from *Agrobacterium tumefaciens*.
A foreign gene is inserted into the Ti plasmid and this is used to infect
plants, producing a crown gall. The plant cells in the gall all contain the Ti
plasmid with its piece of foreign DNA. Gall cells are then grown in culture
to produce plantlets, which can be transferred to soil where they grow into
mature plants. Since each plant is derived from a single cell carrying the
foreign gene, all the cells in the fully grown plant contain that gene.

rapidly. Here, then, is a potential vector for introducing bacterial genes into plants, and the Ti plasmid is being studied in many laboratories, particularly in Ghent and Cologne where its discoverers, Marc van Montagu and Josef Schell, work.

Once a suitable vector has been found, the panoply of genetic engineering tricks described in Chapter 2 become feasible, and one can also attempt to insert genes from one plant into another plant species. A notable feature of this kind of research is the possibility of regenerating a whole plant from a single cell, at least in the case of some plant species. Gardeners are familiar with a very similar process called vegetative propagation or, more commonly, taking a cutting. There is something very remarkable about the fact that, with care and attention, a small piece snipped off an adult plant can grow into a fully fledged replica of its parent. The equivalent operation with animal cells produces quite different results. If the isolated animal cells manage to survive at all (which they often do not), one only obtains a mass of cells which are of the same type as the original cell – for example, hamster kidney cells grow to produce more hamster kidney cells, not whole hamsters.

This phenomenon bears directly on one of the greatest puzzles in biology. Every cell in an organism contains all the DNA needed to specify everything required to build up the whole organism, but in individual animal cells certain genes become irreversibly switched off at an early stage in the animal's development. The genes needed to make paws and whiskers are still present in hamster kidney cells, but they seem to be permanently blocked. In many types of plants, this is not the case; tobacco stem cells can give rise to whole tobacco plants because all their genes are potentially active and, given the right conditions, will produce a complete tobacco plant from just one cell.

This has many important practical consequences, both for plant genetic engineering and conventional plant breeding. In the latter, it means that it is often possible to grow thousands of clones (identical plants) from one parent plant (see page 128).

Figure 2 illustrates how this property may be used by genetic engineers. This has been attempted and, while the results have been mixed, in some cases the trick has worked, and the offspring of the genetically engineered plants also carried the new gene.

Unfortunately, this does not bring the introduction of nitrogen-fixation genes into cereal crops as close as one might think. In fact, few people expect to see nitrogen-fixing cereal crops within the next decade. There are two major obstacles, neither of which should prove insuperable.

Although *Agrobacterium tumefaciens*, with its Ti plasmid, can infect an enormous range of plant species called dicotyledons,* it cannot infect

---

* The terms monocotyledon and dicotyledon refer to the one or two seed leaves, respectively, a species possesses. Many forest trees, such as oak, elm and beech, as

cereals or other monocotyledons. However, there are other potential plant vectors, and the present lack of a suitable virus or plasmid for genetic engineering of cereal crops need not put an end to the long-term aim.

One possibility makes use of two tools of the biotechnologists' art encountered in previous chapters – liposomes and protoplasts. It is possible to package pieces of DNA inside liposomes and it is not too difficult to remove the sturdy cell walls of plant cells to create protoplasts. The general plan would be to wrap *nif* genes into liposomes and mix these with wheat cell protoplasts. With the protective cell wall removed from the plant cells, there is a good chance that the liposomes could deliver the foreign genes into their new home. This has not yet been achieved for *nif* genes and wheat cells, but similar experiments with other gene/plant cell combinations have proved successful. This technique is clearly likely to be of commercial importance if it is possible to grow whole plants from individual genetically engineered protoplasts, and this brings us to the second major challenge in working with cereal plants.

While many sorts of plant can be propagated from small cuttings or single cells, others, including wheat and corn, cannot at present. This problem may be overcome by fusing, for example, a wheat cell carrying the *nif* genes and a cell from a plant which can easily be regenerated from a single cell. The resulting hybrid cell might be very similar to a wheat cell, but would also contain the *nif* genes *and* be able to grow into a complete plant. This proposal is analogous to the methods employed to create hybridomas (page 81).

Obviously there are many imponderables on the way to cereals which make their own fertilizers, but a fair degree of optimism seems warranted when one considers the recent advances in genetic engineering and cell fusion, nearly all of which were science fiction only a decade ago. Indeed, plant genetic engineering offers almost boundless possibilities for farming in the nineties. The following are just a few of the other projects which plant genetic engineers are undertaking.

Despite its apparent simplicity, nitrogen fixation involves some complicated chemical reactions; not only is gaseous nitrogen converted into ammonia, but hydrogen gas is also produced. This is a wasteful process, for there are great quantities of energy locked up in hydrogen which, if tapped, could be put to good use, particularly to fuel the fixation of yet more nitrogen. This would benefit the bacteria's plant hosts since they provide most of the energy to keep the whole machinery in operation.

A survey of *Rhizobium* associated with soybean plants in the US revealed that many of the bacteria, especially in the north and east of the country, contain *h*ydrogen-*up*take (or *hup*) genes. These genes seem to confer the

---

well as fruit trees, potatoes, tomatoes, peas, beans and cabbages, are dicotyledons. Monocotyledons include bananas, grasses, palms, tulips and orchids.

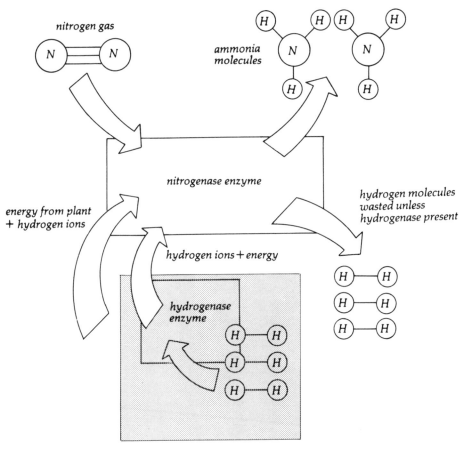

**Figure 3**
The nitrogenase enzymes of bacteria fix nitrogen by converting nitrogen gas
(N₂) into ammonia (NH₃) using energy and hydrogen ions supplied by their
plant hosts. This process also produces energy-rich hydrogen gas. Some
bacteria also have hydrogenase enzymes which can capture hydrogen gas,
converting it into hydrogen ions and releasing energy which can be fed back
into the nitrogenase system.

ability to recycle hydrogen gas back into the nitrogenase system which
fixes the nitrogen, thus harvesting the energy in hydrogen which would
otherwise have been lost to the plant (see Figure 3).

The most straightforward application of this discovery is to introduce
the *hup* genes into those strains of *Rhizobium* that now lack them. The *hup*
genes in certain other types of bacteria are found on plasmids, and if the

same is true of *Rhizobium* the *hup*-carrying plasmids can be transferred from one strain of these bacteria to another. The addition of the ability to utilize the energy in hydrogen gas would not guarantee significantly increased crop yields, since any change in an organism's capabilities is likely to have many ramifications, some beneficial and others possibly not. The sheer complexity of life on the molecular level, with its maze of intricate relationships between one function and all the others (in this case, hydrogen uptake and growth rates), makes it almost impossible to predict with certainty the entire consequences of a single alteration. Here, the problem is compounded by the symbiotic relationship between *Rhizobium* and soybean plants, making it even more difficult to estimate the effects on the two organisms. As usual, experiments will have to be performed to test the effects. If the plants really do grow better, one would aim to introduce *hup* genes directly into crops which have also been given the capacity to fix nitrogen.

Other genes which are being viewed with interest by genetic engineers are the *osm* genes, which are in some way related to a plant's ability to withstand certain stresses, such as lack of water, heat, cold and salty soils. All these adverse conditions have the effect of forcing water into or out of plant cells by the process of osmosis (hence the name of the genes involved in counteracting it). Millions of hectares of land throughout the world are useless for agriculture because of cold, insufficient water or high salinity. A target for the future (at least a decade ahead) is to introduce *osm* genes into crop plants with the aim of opening up vast tracts of barren land to agriculture.

Perhaps nearer at hand is a comparatively simple modification to plant genes which could be of great value to humans. The balance of amino acids in many of the proteins contained in the seeds of cereals – the storage proteins – is not ideal from the point of view of nutrition for humans and domesticated animals. In particular, most storage proteins are deficient in lysine, one of the essential amino acids; for this reason, lysine is often added to animal feedstuffs (see p. 140).

It seems possible that by altering just a few bases in the DNA which codes for these proteins, the plants could be induced to manufacture proteins that better suit our nutritional needs, without doing the plants much harm. As a rule, this kind of tampering with an organism's proteins is likely to be harmful – for example, if one alters the amino-acid sequence of a plant enzyme, its essential catalytic activity will probably be damaged, leading to the death or disablement of the plant. In the case of storage proteins, there is a little more leeway for molecular tinkering. As their name implies, storage proteins are essentially places where amino acids are packed away until they are needed for the construction of other proteins. There is good reason to believe, therefore, that slight modifications to storage proteins – say, by inserting a few extra lysines – would not greatly harm the plant, but might make it more nutritious.

# The ravages of plant diseases

In the mid-eighteen-forties potato blight wiped out a major source of nutrition for the poor inhabitants of much of Europe. Ireland suffered most acutely, losing a third of its population to famine and emigration, as potatoes turned from vigorous plants to withered stems in just a few days. In 1970, an epidemic of southern corn leaf blight swept through the US, destroying maize valued at about $1 billion. Both caused by fungal infections, these are just two of the many dramatic examples of the continuing attrition of our food supplies.

The science of plant pathology – the causes of plant diseases – still lags behind that of human ailments, but over the last decade or so a far better understanding of many plant diseases and what might be done to stop them has been gained. Many economically important plant diseases are remarkably specific. A particular form of the fungus that causes late blight in potatoes may spread like wildfire through a population of one type of potato, but will leave unharmed other, very similar varieties of the plant. An obvious solution may appear to be to plant only the resistant varieties of potato, but, needless to say, nature is rarely that simple. A potato that is resistant to one type of fungal infection may be susceptible to another and, furthermore, that potato may not possess the right taste, shape, colour and all the other factors which make up a marketable product.

Over the centuries, and particularly in the last few decades, plant breeders have had immense success in interbreeding different plant varieties to produce hybrids which possess the good features of both parents. Virtually all the world's major crop plants have been developed by conventional techniques, such as cross pollination. However, this type of plant breeding has an inherent limitation – only quite closely related plant varieties can be interbred to give fertile hybrids. The rapidly developing techniques of protoplast fusion can help overcome this natural obstacle to the creation of new plant varieties. The tough walls that surround plant cells can be removed to produce these naked cells. Protoplasts from quite dissimilar plant species can be induced to fuse together forming a single hybrid cell. In some cases, a complete hybrid plant can be grown from this single cell – for example, 'pomato' plants have been produced which are hybrids between tomatoes and potatoes.

Much effort is now being invested in the search for new crop varieties which are resistant to as many diseases as possible. Tobacco plants are particularly easy to manipulate in the laboratory, and protoplast fusion techniques have already yielded a new variety of tobacco plant which seems to be resistant to hornworm, a prevalent disease in tobacco farming. While research into these methods of developing new crops has only just begun, it has already stimulated considerable investment from food-manufacturing firms – for instance, Campbell's Soups, have put more than $10 million into the search for a better variety of tomato.

The regeneration of plants from single cells taken directly from plant tissues rather than produced by protoplast fusion is, by contrast, of great economic importance today. Asparagus, pineapples, strawberries and oil palms can all be propagated now by separating cells from an especially suitable plant and growing complete replicas of that plant from each individual cell. The same technique is also in general use for the breeding stock of many other species including sprouts, cauliflowers, bananas, carnations and ferns, and even redwood trees are now being tested.

The great advantage of this method of propagating plants is that many thousand identical copies or clones of a *single* plant can be produced. If one simply grew plants from the seeds of a particularly suitable plant, there would be a great degree of variation among the seedlings since each would have a different genetic constitution depending on the way the genes from its *two* parents were distributed. Furthermore, many viral diseases of plants are passed from generation to generation through the seeds. By cloning virus-free plants the spread of these diseases can be reduced.

In economic terms, one of the most significant developments in plant cloning is the introduction of over 1000 cloned oil palms in Malaysia. Palm oil sales are about $2.5 billion a year, including its use in margarine. Once an individual palm was discovered to be unusually resistant to disease and to yield 20–30 per cent more palm oil than the average obviously it made sense to produce multiple copies of this plant. Unilever achieved this in the late seventies, and their test plantation is showing good results.

More recently, researchers in Australia have been able to clone river redgum trees. This is significant because these trees can grow in very salty soils which are unfit for agriculture, and such land could be reclaimed by planting redgums. The hope is that as they grow, they will pump water from the soil, lowering the water table and allowing rain to wash the salt out of the surface layers of the soil. Since about 400,000 sq. km (154,000 sq. miles) of land around the world are severely affected by excess salt, this project has immense potential. Even if the redgums or other salt-tolerant trees prove unable to open the way for conventional agriculture, the wood obtained from otherwise unproductive land would be valuable, particularly in countries, such as India, which have large areas of salt-laden earth and a great need for fuel.

# CHAPTER SIX

# Biotechnology from farm to supermarket

Good health without good food is next to impossible. While the finer points about what constitutes 'good' food are much debated, the plain fact is that many millions of people are chronically malnourished by any reasonable standards.

Historically, biotechnology found its first applications in the production of food – in the making of bread and alcoholic drinks, and in the harvesting of algae as food, practised by the Mexican Indians centuries ago. The present and future applications offer food in more abundance, at a lower cost, higher in nutritional value, greater in variety and more appealing in taste than those available today.

The greatest impact of biotechnology so far has been in the food-processing industry, but the coming years will see important advances in the production of food derived from microbes, plants and animals, which should benefit both the malnourished and the well-fed sections of the world's population.

## Beefing up the meat industry

An outbreak of foot-and-mouth disease is a devastating blow to cattle farmers, entailing the destruction of all affected cattle, which may be worth many thousands of pounds, and all movements of animals in the area must be halted. This disease also affects pigs, sheep and goats, and is endemic in Asia, Africa and South America. Because it spreads so easily, no area of the world is safe from its ravages, and it is the most economically important infectious disease of farm animals.

Foot-and-mouth disease is caused by a small virus related to those that produce polio and the common cold. At present, countries which have only occasional outbreaks (such as Britain and the US) control the disease by slaughtering infected cattle. This is costly, as is the current vaccination process employed in places where the disease is common. Clearly, both areas would benefit greatly from cheaper and more effective vaccines against the disease, and these should become available very soon.

The importance of the problem has attracted many scientists in industry and government establishments, notably the Animal Virus Research Institute at Pirbright, UK and the Plum Island Animal Disease Laboratory, New York State. The gene which codes for one of the virus's coat proteins, VP1, has already been cloned and inserted into *E. coli*. Once purified, this protein is very likely to act as an effective vaccine against the foot-and-mouth disease virus.

Over one billion doses of the present vaccine are injected each year but this vaccine, which is obtained by infecting cultures of hamster cells, is both expensive and must be given annually to maintain a high level of immunity. The VP1 vaccines produced by genetic engineering could offer substantial advantages in terms of cost and effectiveness.

The vital economic importance of eradicating, or at least controlling, viral diseases in animals and the role genetic engineering can play in such endeavours has been recognized by the United Nations Industrial Development Organization. This body has assigned a high priority to the development of genetically engineered vaccines against several diseases, including foot-and-mouth; rabies, which causes economic losses of tens of millions of dollars, especially in South America; African horse sickness, which erupts periodically and, for example, killed 300,000 horses, mules and donkeys in the Middle East in the late nineteen-fifties; and a sheep disease, bluetongue, which is concentrated in Africa but also appears in southern Europe and the US.

Veterinary vaccines are important, not only for their intrinsic value but also because they can be used as proving grounds for technologies that later may produce vaccines for human diseases. The first product of genetic engineering ever to go on sale was a vaccine against a bacterial disease of pigs. Scours kills about 10 per cent of piglets grown in intensive rearing houses in The Netherlands, and in 1982 the Dutch firm Akzo began to market a genetically engineered vaccine against this disease. Similarly, poultry bred in factory farms are susceptible to Newcastle disease virus, and there is a large and valuable market for a vaccine against this illness. Even greater benefits would accrue from a vaccine against animal forms of sleeping sickness; subduing these diseases, which are transmitted by tsetse flies, could open up vast tracts of Africa for the breeding of cattle.

Animal interferons, the natural anti-viral agents, also have the potential to cut farming costs. As with human interferon, research has been severely retarded by the lack of sufficient animal interferon. Recently the supply has been increased by cloning the gene for bovine interferon in yeast cells. The product collected is now being tested to see if it can be used on a wide scale, at an economic cost, to protect against or cure viral diseases in cattle.

As well as keeping farm animals healthy, biotechnology may increase yields very significantly. Injections of extra bovine growth hormone can increase a cow's milk production by up to 40 per cent, and cause cattle to put on 10–15 per cent more weight than normal. Given the enormous importance of the dairy and beef industries, it is not surprising that at least four major genetic engineering companies are aiming for a share of the vast potential market for bovine growth hormone – $250–500 million a year in the US alone. The gene has been cloned and it is likely that the product will become available by the mid-eighties, probably to be followed by its sheep and pig counterparts.

The self-shearing sheep sounds a most unlikely beast, but in a sense it already exists. Sheep wool, like human hair, grows out of cells in the skin and its rate of growth is controlled, in part, by a hormone called epidermal growth factor (EGF). If extra doses of EGF are given to sheep, the wool grows more rapidly and each hair becomes narrower. A dose of EGF, therefore, creates a weakness in the hairs as they are being constructed inside the skin. After a few days, this weak point has emerged from the skin into the visible woolly coat, and at that stage simple brushing snaps off the hairs, yielding a harvest of wool without the need for conventional shearing.

As one might expect, most of the research into this strange phenomenon is taking place in Australia, but whether it can be turned to commercial use depends on at least two problems being solved. First, biotechnologists must clone the EGF gene and so produce large quantities of cheap EGF protein; this should present no insuperable difficulties. The second problem is more mundane, but equally important. As the break-point in the wool takes a few days to emerge from the skin, the wool cannot be plucked off until a few days after the EGF injection. Keeping sheep penned for this time is expensive, but releasing them into their grazing lands would result in the wool being shed on their first encounter with a bush, fence or even each other. The task of collecting wool scattered over hundreds of hectares is probably greater than shearing the sheep in the time-honoured fashion. Obviously the economic realities – cost of EGF plus cost of penning sheep versus cost of conventional shearing – will decide whether this intriguing offshoot of biotechnology becomes a commercial proposition.

# Microbes maketh a meal

The term 'food processing' certainly has an unnatural ring to it and today, when packaging can take precedence over content, its negative connotations are not entirely undeserved. However, processing is the rule rather than the exception, for mere cooking is a type of processing. Of more concern in this section are the very common forms of food processing which require the assistance of microbes – for example, the processing of wheat to make bread, grapes to make wine, barley to make beer, milk to make cheese and yogurt, and vegetables to make pickles. Including substances which are added to food to preserve it, render it more palatable or improve its nutritional value – many of which are derived from microbes – makes the influence of biotechnology on our diets even more evident.

Humans have known how to make use of microbes, particularly yeasts, since Neolithic times, long before people were even aware that microbes existed. The discovery that intoxicating drinks could be manufactured from grapes, rice, barley and other cereals must surely have been accidental – a chance contamination of the juice by a suitable species of yeast, such as those which form the bloom seen on grapes. Over the centuries brewers and wine makers learned how to control this capricious process. Some types of microbes produce a stronger and more appealing drink than others, and these would have been selected for future use. Eventually a taste for particular types of drinks became established, and one aspect of biotechnological development – the selection of the most appropriate microbes – declined in importance.

The use of different strains of yeast will produce changes in, for example, a beer, and any changes must usually be kept to a minimum once the product has become accepted by the consumers. The brewing industry would like a yeast which produced a beer higher in alcoholic content since this would reduce processing costs. However, the flavour and appearance of beer depend on many factors, some of which are only poorly understood. It is difficult, therefore, to find a new strain of yeast which possesses desirable qualities, such as making a stronger product, while retaining the correct blend of flavours and other characteristics. For these and other reasons, two types of yeast dominate in the brewing industry. *Saccharomyces cerevisiae* is employed to manufacture traditional British-style beers, termed top fermentation beers because this type of yeast tends to float on the surface of the brewing vessel. Lager beers, by contrast, are made with the aid of a bottom fermentation yeast, *Saccharomyces carlsbergensis*, so-named because it was first isolated in the Carlsberg Institute, Copenhagen, by Emil Christian Hansen.

The brewing industry has laid the foundations of much of the new biotechnological revolution and is the largest biotechnology, with an annual turnover of billions of pounds. It is ironic, therefore, that it stands to benefit rather little from the most recent developments. There are however some possibilities, and one of particular interest concerns light beers. These are beers which have a low concentration of carbohydrates, not to be confused with the more familiar light ales which have been available in the UK for many years. In the US, and to a lesser extent in the UK, light beers have gained much popularity. The brewers' usual yeasts cannot ferment substances called dextrins, which make up about a quarter of the carbohydrates in the original fermentation liquid. A related yeast (which is not used in the industry), *Saccharomyces diasticus*, can turn dextrins into alcohol, thus reducing the carbohydrate content of the beer. The genes responsible for this have been introduced into normal brewing yeasts. At present this work is still at the research stage, and so far the taste of the beer has been far from satisfactory, but further investigations are beginning to show how this may be improved.

The role of genetic engineering in the brewing industry of the future may be uncertain but another well-established aspect of biotechnology has already become firmly entrenched – the use of various enzymes obtained from microbes. One of the major costs in beer production is the malting process during which barley seeds are allowed to germinate. The main food store inside barley consists of starch molecules, which are composed of many sugar subunits linked to form a branching chain (see Figure 1). The sugar units by themselves or linked in pairs make an excellent source of energy for yeasts, and when consumed give rise to alcohol. Brewing yeasts cannot make use of the complex starch molecules but, during the germination of barley seeds, enzymes are released which snip up the starch chains into pieces that yeasts can consume.

Over recent years, a means of avoiding some of the expense of malting has been developed. Certain species of bacteria also manufacture some of the enzymes which attack starch, notably amylases. These amylases are now extracted on a commercial scale from bacteria, and this market is

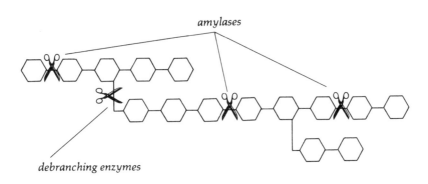

**Figure 1**
Starch consists of branched chains of sugar subunits. Amylase enzymes produced by germinating barley seeds or bacteria are able to snip the links to yield shorter chains of sugars which yeasts can consume. 'Debranching' enzymes cut the branching points of the chains.

worth several million pounds a year. When added to the fermentation vessel, these enzymes can perform some of the tasks traditionally undertaken by the malted barley enzymes, allowing a proportion of the cheaper, unmalted barley to be employed without affecting the final product.

Clarity, as all beer advertisers know, is a strong selling point. There are several reasons why a beer may become hazy, one of which – excess protein in the liquid – can be eliminated with the aid of enzymes. Proteases are, as their name suggests, enzymes which destroy proteins. Unlike most enzymes, proteases are generally quite undiscriminating – present them with any type of protein and they will quickly chop it into small pieces.

The protease most widely used is papain, an enzyme extracted from the papaya fruit (and also employed to tenderize meat by attacking the tough fibres). The Japanese, who are the world leaders in the industrial applications of enzymes, have been investigating microbes which produce proteases. The Kirin Brewing Company has discovered that a mixture of papain and a protease obtained from the bacterium *Serratia marcesens* is more effective in eliminating beer haze than either enzyme alone. Bacterial proteases for these and other applications are sold in vast quantities, over 500 tonnes a year with a value of £60 million worldwide.

Brewers and associated researchers, particularly since the time of Pasteur, have laid a solid foundation for much of biotechnology's future. They have gained unparalleled practical experience over many years in putting microbes to work in an industrial environment – making sure that they are given the right conditions to perform their assigned task, excluding contaminating organisms, dealing with vast quantities of raw materials, refining the products of a biotechnological process to present them in a saleable form, and keeping costs to a minimum. All this knowledge is proving invaluable in transforming newer, and perhaps more glamorous biotechnologies into economic realities.

Second only to the alcoholic beverages industry in terms of sales are the industries which employ microbes to manufacture foods from dairy products, chiefly cheeses, buttermilks, sourmilks and yogurts.

The principles of cheese-making vary comparatively little from one cheese to another, but the specific techniques, particularly the enormous range of microbes employed, differ greatly. The first step is to add bacteria to the fresh milk, allowing them to turn it sour as the milk's lactose sugar is converted into lactic acid. An enzyme which breaks down proteins is then introduced, causing the solids to coagulate before they are separated from the remaining liquids and allowed to mature to form a cheese. At the start of the maturing process, other types of microbe may be added to impart particular characteristics to each type of cheese.

The bacteria which first act on the milk are usually members of the *Streptococcus* group, but there is much more variation in the microbes employed during the later stages. The applications of some of these microbes is betrayed by the names, *Penicillium camemberti* and *Penicillium roqueforti*, for example. Some cheeses, such as Gruyère, have holes inside the body of the cheese which are created by adding to the milk bacteria that give off carbon dioxide gas. Much the same principle applies in the use of carbon-dioxide-generating yeasts to leaven bread.

One of the cheese industry's biggest problems is the occasional failure of the bacterial starter cultures to initiate the fermentation of the milk. In Australia, for example, thousands of dollars worth of milk is wasted by this difficulty, but scientists in Melbourne foresee a solution. Certain types of viruses (bacteriophages) infect and kill bacterial cells. Although some species of *Streptococcus* bacteria are resistant to bacteriophage attack, the strains which are best for cheese-making are susceptible to the viruses. By employing genetic engineering techniques or cell fusion, these scientists hope to create hybrid bacteria which are both efficient milk fermenters and able to fight off bacteriophage infections.

Traditionally, the protein-degrading enzyme employed in cheese-making has been rennin, which is obtained from the stomachs of calves. Both the cost and potential shortage of calf rennin stimulated investigations into alternative enzymes obtained from microbes. Over the past fifteen years or so, microbial rennin has become increasingly widespread in the cheese-making industry. Three species of fungi, *Mucor pusillus* (developed in Japan), *Mucor miehei* (US and Denmark) and *Endothia parasitica* (US) make rennins which have been approved for use in most countries. This is another traditional area of biotechnology that has tempted the genetic engineers, who hope to reduce the cost of these enzymes. At least two groups of researchers, in the UK and Japan, have managed to clone the calf rennin gene in bacteria and yeasts.

Other dairy products (buttermilk, sourmilk, yogurt and so on) are made by mixing specific types of microbes with fresh or pre-treated milk. Vegetables may be preserved by fermenting them with bacteria that produce acids to yield foods such as sauerkraut. Even the kind of pickling done in the home depends indirectly on the action of microbes; vinegar is made when certain microbes feed on grape juice or infusions of cereal grains. Civilizations in the Far East have been even more adventurous in the use of microbes. In Japan, China and Indonesia, delicacies such as natto, sufu and tempeh kedele are made by growing moulds on soybean preparations, and the ubiquitous soy sauce is manufactured with the aid of several fungi, bacteria and yeasts.

# Bug grub

It is not only in preserving and altering the tastes of foods that microbes have a role to play – microbes themselves can be eaten. The prospect of trying to sell a food consisting of little more than squashed, dried microbes or 'bugs' is far from appealing to marketing consultants. To avoid public revulsion – unjustified though it may be – the more wholesome-sounding, but less accurate, term single-cell protein (SCP) has been adopted.

The origins of human consumption of microbes can be traced at least as far back as the early sixteenth century. In 1521, Bernal Diaz del Castillo described how the inhabitants of Mexico ate small cakes with a cheese-like flavour, which they prepared from an 'ooze' extracted from lakes. This 'ooze' was almost certainly *Spirulina maxima*, a species of alga that still thrives in the highly alkaline waters of Lake Texcoco. It is not known quite when the Aztecs discovered the nutritional value of this alga, nor, indeed, how long the Kanembu people of Chad, thousands of kilometres away in Africa, have been eating a related microbe, *Spirulina platensis* (see Plate 9). It is clear, however, that human interest in microbial food far antedated the intensive research it has received in the last two decades.

The modern story of SCP, which may contain up to 70 per cent protein by weight, began in Germany during the First World War. Scientists and technologists in Berlin devised a process by which brewers' yeasts were grown in large quantities and added to sausages and soups to supplement the nation's diminished food resources. Eventually the yeasts made up more than half of the country's restricted food imports. Again in the Second World War, Germany turned to yeasts, this time employing *Candida* species. In Britain, brewers' yeasts have made a contribution to the diet for many years, the excess microbes being sold as animal food or processed to form products such as Bovril and Marmite.

The next major events in the SCP story – which have since been called grandiose follies – took place in the nineteen-sixties. Swept along by a tide of enthusiasm, many of the world's largest firms, especially those in the oil industry, invested heavily in SCP production. The results were little short of disastrous – millions of pounds disappeared into ambitious projects, sometimes without a penny in return.

Of all the foods which are in short supply in the world the most acute deficiency is in high-quality protein foods. Protein in animal meat is generally more nutritious than plant proteins, but feeding plants to animals is a very inefficient way of enhancing protein quality: 100kg of corn protein fed to cattle yields less than a third of that weight of beef, and most of the meat consists of water. If the carbohydrates in corn were used as a feedstock for microbes about 100kg of microbial protein could be obtained in *addition* to the plant protein.

Obviously SCP production makes good sense, but despite this the recent SCP ventures were dismal failures. The reason for this was that, inconceivable though it may seem to us today, accustomed as we are to high petrol costs, some firms decided that there was an economic future in feeding *oil-based* chemicals to microbes which would then be harvested and sold as animal, or even human, food.

British Petroleum, for example, spent tens of millions of pounds developing a factory in Sardinia, which was to manufacture Toprina, an SCP product based on yeasts grown on oil-derived nutrients. Today the plant lies idle and is up for sale with a price tag of only £10 million. The price of oil rocketed, there were political difficulties within Italy, and the company could not convince the authorities that Toprina was safe to use as animal food. A whole chain of objections were put up, many of which impartial observers considered spurious. One of the main concerns over safety centred on the relatively high concentrations of nucleic acids (DNA and RNA) in Toprina. Although every cell in our bodies contains nucleic acids – and human nucleic acids are chemically virtually identical to those found in microbes – a question mark hangs over the safe concentration of these substances in food. Yeast SCP has about 10 per cent by weight of nucleic acids, compared to only 1–2 per cent in plants and 4 per cent in *Spirulina*.

ICI has persevered with SCP production. The company's plant at Billingham grows the bacterium *Methylophilus methylotrophus* on methanol derived from natural gas (see Plate 10). Current production of Pruteen is over 50,000 tonnes a year and it is sold as animal feed. The process now faces competition from Hoechst in West Germany, who have a similar product, Probion, from which much of the nucleic acids and other materials have been removed, and it might eventually be approved as a human food. ICI aims to fight back by reducing the cost of Pruteen with the use of a genetically engineered modification of their current bacterium, one which will make better use of the ammonia that must be supplied to the fermentation vessel to provide the nitrogen essential for microbial growth. The genetically engineered bacterium uses about 7 per cent less methanol, but it has yet to be introduced into commercial operations.

Also in Britain, Rank Hovis McDougall has spent about £30 million in manufacturing an SCP product, mycoprotein, for *human* consumption. This company is investigating a species of *Fusarium* mould which contains about 45 per cent protein and 13 per cent fat, a combination quite as nutritious as many meats. Moreover, the mycoprotein is high in fibrous content, a feature which is viewed with ever-increasing approval by most nutritionists. *Fusarium* grows on a wide variety of carbohydrate sources, so the precise carbohydrate nutrient supplied to the mould could vary according to local conditions – potatoes or wheat starch in Britain, and cassava or other tropical plants where these are plentiful. This product has great potential, particularly since it contains less than the acceptable maximum amount of nucleic acids and has a well-balanced amino acid composition. Several hundred people, both inside and outside the company, have eaten this new food and their reactions have been favourable. Extensive tests with animals have revealed no harmful effects of a diet containing mycoprotein. The use of raw materials which are themselves perfectly edible, helps promote confidence that the product will be safe.

It might be imagined that the cheapest feedstock would be some material that comes free of charge, but even this can be improved upon. Waste and pollution disposal is big business, and biotechnology can convert industries which are paid to eliminate some noxious substance into industries which also utilize these as raw materials to produce valuable products such as fuels (page 151) or foodstuffs.

The paper-making industry faces a serious problem in disposing of one of its waste products, sulphite liquor. If sulphite liquor is simply pumped into rivers and lakes, it quickly depletes the water's oxygen reserves with dire consequences to the environment, the death of thousands of fish being only the most obvious effect. In Finland, sulphite liquor is fed to *Paecilomyces* moulds in what is known as the Pekilo process. This not only purifies the waste liquids from paper factories but also yields a rich harvest of microbes which are sold as animal feed. Similar techniques are being developed throughout the world to utilize waste materials from forestry, cheese-making, unwanted fruit pulps and many other materials for which there are few, if any, present uses.

As the oil-based processes proved all too vividly, the economics of SCP production depend critically on the costs of raw materials and the price of conventional alternatives, such as soybean and fish-meal. The USSR, one of the world's major producers of SCP, already has nearly 100 installations, and expects to have enough yeast SCP to fulfil all its needs for protein in animal feeds by the end of the next decade.

The acceptability of SCP as human food is still largely untested. After hesitantly tasting a sample of bacterial SCP, a nutritionist is reported to have remarked, 'Yes, it has all the characteristics I look for in a new human food; it is odourless, colourless, textureless and tasteless.' This unappetizing description means, of course, that it could be employed as the basis of a wide range of acceptable styles and tastes of food. The idea will not please food purists, but the ever-worsening food crisis may force us to abandon such niceties. Once each form of SCP has passed the relevant safety tests and been constituted into an appealing and preferably familiar form, there is little reason why 'bug grub' (under a suitably attractive brand name) should not play an even greater part in improving the world's diet.

The SCP projects developed in the West, including those already mentioned and the Swedish Symba system which utilizes potato wastes, all require heavy capital investment and fairly sophisticated installations. However, the major protein shortages are found in countries where these resources are scarce. If SCP production takes off in the West, the developing world may well benefit through a knock-on effect, in which more of the well-nourished countries' food can be diverted to those who need it. Indigenous SCP programmes would free protein-starved nations from the vagaries of international politics and economics.

*Spirulina* algae offer excellent prospects for low-technology SCP production which could be employed in these parts of the world. The yields of *Spirulina* per hectare may be ten times higher than wheat in terms of total weight and many more times greater in protein content. Harvesting the algae from natural or artificial lakes is very simple, and it can be dried in the sun and then flavoured according to local tastes. This corkscrew-shaped organism is photosynthetic and so gains its energy from sunlight, using it to build up compounds from carbon dioxide in the air. Since both light and air are free, and only a few cheap chemicals need be added to the ponds to keep productivity high, *Spirulina* has many enticing characteristics.

# Vitamins and amino acids

Biotechnology is also contributing substantially to the improvement of the nutritional value of existing foods. The value of vitamins scarcely needs emphasizing, for constant injunctions from advertisers have ensured that very few readers will be unaware of the severe consequences of too little of them in the diet. Indeed, a more important task is to counter the idea that twice the recommended dose of a particular vitamin must be twice as health-giving. Few healthy people in developed countries, eating a normal diet, need extra vitamins. This happy state of affairs is largely due to their ready access to many vitamin-rich foods, and biotechnology has also had an influence.

A lack of vitamin B12 (cobalbumin), which may be caused by intestinal disorders or, occasionally, an inadequate diet, results in pernicious anaemia. Too little vitamin B2 (riboflavin) can produce lip sores, mouth ulcers, skin rashes and eye problems. Today, much of the supplies of these two vitamins, used in medicine and as food additives, comes from microbes. The interest in these two substances here lies in the great success of biotechnologists in persuading certain species of microbes to increase their production of these vitamins to a remarkable degree. Wild strains of the mould *Ashbya gossypii* manufacture only minute amounts of vitamin B2, but by successive selection of high-yielding strains and manipulation of the fermentation conditions, the quantity of vitamin produced has been increased 20,000 times. Similarly, the industrial strains of two types of bacteria, *Propionibacterium shermanii* and *Pseudomonas denitrificans*, make over 50,000 times more vitamin B12 than their natural cousins. This achievement of biotechnologists has enabled their microbes to form the basis of very efficient industrial processes, supplying most of the $150 million market for these vitamins.

139

Cereal grains constitute a major part of the feed for many economically important animals, especially during the winter. Unfortunately, many cereals are deficient in two of the amino acids, lysine and methionine, which all animals need to build up their proteins. These are, therefore, usually added to animal feed to ensure that an adequate diet is provided.

At present methionine is manufactured by chemical processes, but 80 per cent of the $200 million sales of lysine is fulfilled by fermentation processes which use bacteria. The ever-growing demand for animal feed has encouraged several firms to invest in larger and more efficient facilities for producing lysine, and biotechnological methods of manufacturing methionine are being sought.

About 40,000 tonnes of lysine are produced each year, chiefly by over-producing strains of *Corynebacterium glutamicum*. As explained on page 000, these bacteria lack the enzyme homoserine dehydrogenase, and this defect causes them to continue to churn out vast quantities of lysine – far in excess of their needs.

While several other amino acids are currently being manufactured by microbial fermentation processes, one dwarfs all the others in economic importance. A quick examination of the labels on food packages in the average kitchen will reveal the ubiquity of monosodium glutamate (MSG), which is used as a flavour enhancer. The MSG industry now produces over half a million tonnes a year, and its growth has been based on two bacteria – another strain of *Corynebacterium glutamicum*; and *Brevibacterium flavum*, a short rod-shaped organism.

The cost of MSG is relatively low because biotechnologists have discovered a particularly neat way of extracting it from bacteria. *Corynebacterium glutamicum*, like many other microbes employed in fermentation processes, is provided with the sugar glucose as its main source of carbon and energy. It also needs small quantities of the vitamin biotin, which is used to help build up the membrane that surrounds the cell. If the bacteria are supplied with just too little biotin, their membranes become 'leaky', and some substances, including MSG, escape from the cells into the surrounding liquids from which they can easily be isolated. This cuts down the cost of purifying the product, since it is not necessary to break open the bacterial cells to collect the MSG. Furthermore, the amount of MSG manufactured by each bacterium increases because it then replaces the MSG it has lost.

# Sweeter than sweet – sugary substitutes

The commonest disease in Britain is caries, or tooth decay. Virtually every adult has one or more fillings, and millions have lost all their natural teeth. Sugar, which is added to thousands of our foods, is thought to be the major culprit, nourishing billions of bacteria in our mouths which promptly produce acids to corrode teeth and gums. The enormous scale of the sugar industry has made it a prime target for biotechnologists. Already several alternative sweeteners are available through the skills of microbes, and some of these alternatives offer the prospect of reduced tooth decay.

Chemists use the term sugar to describe a great number of compounds which share certain physical and chemical characteristics, but only a small number of sugars taste sweet. The most familiar of the sweet sugars is sucrose, which is obtained from sugar cane and sugar beet. A similar compound, fructose, is found in fruits and honey. The biotechnological production of fructose is a major and rapidly growing industry, which is perceived as such a threat to sugar beet farmers that the EEC has blocked its use in the Community. While the use of fructose does not promise better dental health since plenty of acid-producing bacteria can feed on it just as well as on sucrose it has attracted great attention for three main reasons: it is about twice as sweet as sucrose; it can prove cheaper than sucrose, especially since the price of cane and beet sugar fluctuates rapidly and unpredictably; and fructose is more suitable for diabetics than ordinary household sugar. In the years since 1970 the use of fructose sweeteners in the US has risen from almost zero to nearly 20 per cent of the sweetener consumption of about 50kg (110lb) per person per year, and before long half of the sweetening in Coca Cola may be provided by this sugar. Fructose tablets were briefly in vogue in Britain when the alcohol breath test was introduced since there were ill-founded claims that they reduced blood alcohol concentrations.

The remarkable impact of fructose has been based largely on advances in biotechnology, in particular the availability of amylase enzymes which convert cheap starch raw materials into glucose, and another enzyme, glucose isomerase, which turns glucose into fructose. The principles are familiar: the impressive catalytic powers of enzymes are harnessed to provide an economic and efficient means of changing a raw material of low value (starch) into one of higher value (fructose). There is one aspect of these processes that deserves special attention since it has considerable importance in many diverse areas of biotechnology – the immobilization of enzymes.

Since 1928 when James Sumner became the first person to obtain pure crystals of an enzyme – urease from jack-beans – scientists' ability to isolate specific enzymes has increased beyond measure. However, they remain, in general, rather expensive materials and so it is vital that any commercial process which employs enzymes should utilize them as efficiently as possible. One of the characteristics of enzymes is that they can perform the same function over and over again; barring accidental damage, a single molecule of the enzyme glucose isomerase will continue to work tirelessly converting glucose molecule after glucose molecule into fructose. Once all the glucose has been turned into fructose the enzyme has nothing left to work with and its undiminished vigour is wasted. Since the enzyme and the fructose are both dissolved in the final mixture, it is very difficult and costly to extract the enzyme and use it again with a fresh batch of glucose.

This waste of expensive enzyme can now be avoided by the simple expedient of fixing the enzyme molecules on to a solid surface; once they have done their job, the fructose is drained away leaving the enzymes stuck on their supports ready for the next infusion of glucose. There has been much research into the theory and practical applications of immobilized enzymes over the last decade, and many different types of enzyme have now been successfully immobilized on a wide variety of solid supports including glass beads, plastics and natural fibres, such as cellulose. Figure 2 illustrates three of the main methods employed to fix enzymes.

In early experiments, the activity of enzymes was frequently destroyed during the immobilization process. Now the techniques have been refined to such a degree that nearly all the enzyme molecules will be unharmed when they are linked to the supports. These advances, coupled with the expectation that it will be possible, eventually, to immobilize any desired enzyme, guarantees an increasingly important place for immobilized enzymes in biotechnology's future. At present about fifty microbial enzymes are of industrial importance, but patents have been filed for commercial uses of over 1000 enzymes. There are many and varied reasons why the majority of this large number of potential applications have not been brought to fruition. The high cost of enzymes is certainly a major factor in many cases but immobilized enzyme technology is transforming many industrial processes. The re-utilization of a single batch of enzymes can turn a previously uneconomic biotechnological process into one with great potential. This has happened in the production of fructose, of the food additive malic acid, and of several amino acids.

Although fructose production is the most significant contribution of biotechnology to the sweetener industry at present, at least two other types of sweeteners will probably become important in the next few years. One of these, a synthetic compound called aspartame, has recently been approved for use in the US and the UK as a low-calorie sweetener. The other, which consists of two related materials, thaumatin and monellin, is being investigated intensively.

Aspartame is quite a simple material composed of two linked amino acids, aspartic acid and phenylalanine. Both the constituents can be manufactured by microbes and aspartame itself will probably soon be produced by a microbial fermentation process. This market is worth over $4 million in the US and rising rapidly, so there is a large incentive for biotechnologists to break into it.

Thaumatin and monellin are much larger compounds, consisting of 207 and 94 amino acids, respectively. Both proteins are remarkably sweet – about 100,000 times as sweet as cane or beet sugar. Obviously this means

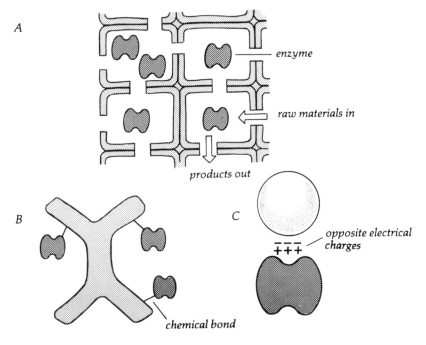

**Figure 2**
Some methods of immobilizing enzymes. (A) Entrapment in a solid lattice
constructed of plastic or other materials. Small gaps in the lattice allow the
raw materials to approach the enzymes and the products to leave the
compartments. (B) Chemical bonding of enzyme to a support, such as
cellulose fibres. (C) Electrostatic attachment to supports such as glass
beads. The opposite electrical charges possessed by the enzyme and the
support hold the two together.

that tiny quantities would satisfy the tastes of even the most sweet-toothed
consumers. They are both found in west African bushes, but the cost of
extracting these materials from plants is high and is likely to limit their use
in the food industry. Recently the gene for thaumatin has been cloned and
there are now excellent prospects that genetically engineered microbes will
manufacture it at a significantly lower price.

# CHAPTER SEVEN

# Bioenergy and fuels for the future

Amid the clamorous warnings of an impending energy crisis, it is easy to overlook the fact that there is enough energy in our surroundings to meet just about every conceivable need. Despite this fact, the concern about the future of our energy supplies is not a figment of a few pessimistic minds. The crux of the problem is that we do not yet have the right technologies to tap much of this energy.

The great majority of our energy resources are ultimately derived from the sun, including wood, coal, oil, natural gas and even wind and hydroelectric power. Truly vast amounts of energy could be obtained from this source: the energy in one day's sunlight is equivalent to about a fifth of all the known reserves of fossil fuels. To plug the energy gap, we must devise ways of converting this plentiful supply into forms of energy that are convenient to transport, store and consume. Usually this means producing liquid, solid or gaseous fuels, which may be used directly or converted into yet another form of energy, electricity, for example. It is now widely recognized that we need to wean ourselves from dependence on fossil fuels, such as oil, coal and natural gas, and to build a more stable energy economy based on renewable and less polluting fuels. This is where the biotechnologists come in, for the skills of microbes and plants can be used to produce just such fuels. Some energy biotechnologies, through the production of fuel alcohol and methane gas, are already making a significant contribution to the energy budget. Others, such as the production of hydrogen gas, are not yet operating on an industrial scale, but clearly offer great promise.

## Alcohol and energy from sugar

Millions of tiny yeast cells, thousands of hectares of sugar cane plants and more than a quarter of a million Brazilian motorists are linked by one of the fastest growing biotechnologies of the eighties – the production of fuel alcohol. Instead of filling their tanks with petrol, drivers throughout Brazil are powering their cars with alcohol derived from sugar canes. It costs only

£200–300 to modify an ordinary petrol engine so that it can burn alcohol efficiently, and with fuel alcohol on sale at little more than half the price of petrol, there is a good incentive for Brazilians to make this investment. Indeed, some of the cost benefits of fuel alcohol can be obtained without altering the engines at all; they can be powered with a mixture of petrol and alcohol known as gasohol (plate 11).

Many countries are now following Brazil's lead. Ambitious projects aimed at substituting alcohol for petrol are under way in the US and the EEC, and fuel-alcohol installations are to be found in countries as diverse as Zambia, the Philippines, Zimbabwe, Nicaragua and Paraguay. Gasohol, consisting of 10 per cent alcohol and 90 per cent petrol, is now widely available in the US. The scope of this biotechnology is indicated by the declared aim of the US Congress that a tenth of all petrol consumption in that country should be replaced by alcohol before the end of this decade; this will entail producing over 50 billion litres (11 billion gal.) of fuel alcohol each year. If these bold targets are met and vast quantities of fuel alcohol become available at a reasonable cost, it will be due largely to the current efforts of biotechnologists.

There are three intertwined strands to this research. First, the chemical reactions by which yeast cells ferment sugar to form alcohol are being investigated in great detail. Second, there is a search for cheap and abundant raw materials to replace or supplement those now fed into fuel-alcohol factories. Third, the best fermentation conditions must be discovered, so that as much alcohol as possible is obtained from the raw materials.

### How and why do yeasts make alcohol?
There is a fundamental aspect of cells which has been touched upon only obliquely – their activity as energy converters. No industrial process is as versatile and efficient as a living cell in extracting energy from the available resources and converting that energy into a useful form. But what is *useful* energy for a cell?

Almost every activity of a cell requires energy – energy to move, energy to synthesize compounds, energy to grow. Some parts of a cell's metabolism are chiefly devoted to building up the compounds the cell needs, and other parts break down food molecules to supply the energy that keeps the manufacturing processes going. Although the nutrients reaching the cell contain a great deal of energy this cannot be used directly, but must first be extracted from the nutrients and stored in a few energy-rich compounds, pre-eminent among which is ATP (adenosine triphosphate).

ATP can be regarded as the cell's energy currency. Energy enters the cell in a variety of forms (nutrients) and the cell's energy metabolism acts like a *bureau de change*, converting foreign into local currency (mostly ATP). Appropriately enough, there is a commission to be paid at this stage – some energy is always lost during the conversion. The local currency can then be deposited and later spent anywhere in the cell to buy (make) the materials the cell requires.

Clearly, then, it is of prime importance to a yeast cell that it maintains its stocks of ATP, and it can do so by consuming sugars. Among the best-known sugars are glucose and fructose. Molecules of both these sugars are found in fairly high concentrations in fruits, but in cells they are usually combined with other molecules to form more complex materials. Sugars have the very important property that they can join together in chains. For example, sucrose consists of just one glucose molecule attached to one molecule of fructose. Sucrose is the main component of sugar cane juice which is crystallized to form household sugar. The yeast's first step in obtaining energy from sucrose is to break apart the glucose and fructose units. These are then fed into the energy metaboliz-ing machinery to provide energy that the cell can use. If the yeast cell has access to a good supply of oxygen these sugars are broken down, step by step, into smaller and smaller molecules. This process extracts the maximum amount of energy from the sugars and at the end only carbon dioxide and water remain, which are discarded. However, if there is little or no oxygen available to the yeast the series of chemical degradations cannot be completed and the sugars are broken down to form molecules of ethanol. Ethanol is the most familiar of the many different types of alcohol. Indeed it is often referred to simply as alcohol or, in the present context, fuel alcohol.

Since alcohol is the product sought, biotechnologists – like brewers from time immemorial – focus their efforts on persuading the yeasts to make as much of it as possible. It is important to note that, so far as yeast is concerned, alcohol is a waste product; it simply accumulates because there is not enough oxygen available to the cell to release the energy from alcohol. Indeed, alcohol is not merely useless to yeast under these conditions, it is positively harmful. When the concentration of alcohol in the yeast's surroundings reaches a certain level – usually 12–15 per cent – the cells stop growing and may even die.

This fact has important consequences for fuel-alcohol factories. A significant proportion of the total production cost arises from the need to purify the alcohol from the mixture of materials in the fermentation vessel. Normally the mixture is heated so that the alcohol boils, and the vapour is then condensed to give fairly pure alcohol. This distillation process requires a great deal of heat. If the fermentation mixture contained 24 per cent alcohol instead of 12 per cent the distillation costs would be reduced considerably.

More is known about the biochemistry and genetics of yeasts than about any other microbe except E. coli. Despite this, there is a paucity of hard facts to explain why some yeasts are more tolerant to alcohol than others, and this makes it difficult to conduct a rational search for the tolerant strains.

One idea is that more tolerant strains have slightly different types of membranes. The membranes which surround cells contain large amounts of the fatty compounds known as lipids. Supplying the yeasts with different lipids may result in them being built into the yeasts' membranes and conferring a greater resilience against the onslaught of alcohol. Even if this approach is unsuccessful, the conventional trial-and-error searches will probably unearth yeasts that are better suited to fuel-alcohol production.

Because biotechnologists need not worry about the palatability of fuel alcohol, they can use a wider range of raw materials and microbes for its production than the few that give an acceptable alcoholic drink. Against these advantages must be set the crucial consideration that fuel alcohol must be cheap, for any biotechnology will only come to commercial fruition if it can compete in price with alternative methods of manufacturing the same product. The biotechnological production of fuel alcohol is an example of price competition at its fiercest – it must not only compete with other types of fuel, but must also face the challenge of alcohol manufactured by purely chemical processes, already a well-established industry.

## Cutting the cost of raw materials

The Brazilian National Alcohol Program is producing over 4 billion litres (nearly 1 billion gal.) of alcohol a year, mainly from sugar cane juice, while corn (maize) is the favoured raw material in the US. Both of these crops have the drawback that they have a relatively high value as food for human or animal consumption, and so their price on world markets fluctuates rapidly and unpredictably according to the annual harvest. Consequently, it is difficult to forecast whether an investment in fuel-alcohol factories based on these raw materials will be profitable. Much effort is being put into the search for alternatives that have lower and more stable costs.

A wide range of plants are receiving close attention including cassava – a root crop that grows on poor soil throughout the tropics – potatoes, trees and grasses. None is as immediately attractive as sugar cane because they do not contain significant quantities of the sugars that yeasts can consume (see page 133). However, this problem is far from insurmountable. Like all plants, they are rich in polysaccharides, and polysaccharide molecules simply consist of many sugar molecules joined together. Unlinking the sugar units will provide an excellent food which yeasts can convert into alcohol.

Two types of polysaccharide are of particular interest, cellulose and starch. Cellulose molecules consist of anything between a few hundred and several thousand glucose units linked together in a straight chain. Starch is also made up of glucose units, but there are fewer in each molecule and they form branched rather than straight chains. Many plants store vast amounts of starch, and cellulose is a major component of all plants, providing the strength of the walls that surround their cells. It has been estimated that over a billion tonnes of starch is formed each year by the world's cereal and root crops, while the figure for cellulose is much higher; indeed cellulose is the most abundant organic material on Earth.

Many existing fuel-alcohol factories use starch as the basic raw material, particularly in the US where over 2.2 billion litres (500 million gal.) of alcohol will soon be produced from corn. A number of chemical and physical processes are involved in treating plants to yield sugar. An important preliminary is wet-milling, which extracts the starch from corn. Much of this book has extolled the ability of enzymes to perform all manner of chemical reaction and it should come as no surprise, therefore, to find that enzymes can produce sugar from starch. Most yeasts do not have the right enzymes, but other organisms do.

Plants use starch as an energy store, and when they need to mobilize their reserves they call upon the services of enzymes called amylases which break down starch molecules, freeing the sugars. They are among the most effective enzymes yet discovered – each molecule can participate in up to 1000 separate chemical reactions every second. Amylases can be extracted from certain species of bacteria and added directly to the starchy raw materials, where they perform the necessary transformation very efficiently. This process will undoubtedly be further improved when immobilized enzymes come into general use.

The ideal microbe would be one that has both amylases *and* the ability to produce alcohol. In fact, some yeasts do possess amylases, but unfortunately they manufacture little alcohol. There are two possible ways to circumvent this obstacle: widen the search for microbial recruits to the fuel-alcohol industry; or use genetic engineering to create a yeast that can do what is required. A few years ago the latter would have been quite unthinkable, but now it looks very possible. The gene that tells bacteria how to make amylase appears to be uncomplicated, and so there is little reason why it cannot be snipped out of a bacterium and inserted into yeast cells, which will then start producing amylase. If the gene were introduced into the most highly productive strains of yeasts now available, the result would be vast quantities of alcohol from starch without all the expensive pretreatment.

However, the new techniques of genetic engineering offer a yet more enticing prospect – yeasts that can feed on cellulose. Most people are all too familiar with the ability of some fungi to attack cellulose. Dry rot, for example, is a fungus that makes enzymes known as cellulases which decompose the cellulose in the cell walls of timber and produce glucose in the process. A number of exotically named fungi – including *Trichoderma reesei*, *Coriolus hirsuitus*, and *Polyporus anceps* – can perform this trick. The insertion of cellulase genes into yeasts would pay enormous dividends. The fuel-alcohol industry could then draw on a huge supply of plant material that is currently untapped. This is particularly important since most of the starch-containing plants (such as potatoes, cassava and corn) are needed as food for humans or animals. A fuel-alcohol industry based on cellulose could make use of almost any plant material – trees, weeds, scrub, straw – and industrial wastes, such as those emanating from paper-making factories.

A drawback is that the cellulose in woody plants is intimately associated with lignin, a compound that is remarkably resistant to degradation. However, some fungi can attack lignin, and when more has been learned about the way they perform this feat, it may be possible to create yeasts which can do it as well. In the meantime, biotechnologists are developing two-stage processes: fungi are used to break down cellulose and lignin to produce sugars, which are then fed to yeasts for fermentation to alcohol. Studies in several countries, particularly Sweden and the US, are revealing the great promise of this approach.

The prospects of enhancing the efficiency of fuel-alcohol systems are not entirely dependent on schemes for broadening the range of raw materials that are used. Increasing the speed and efficiency with which a given amount of sugar is converted into alcohol is equally important, and this provides an excellent example of the way that many years of skilled pure research may yield unexpected practical benefits.

Yeasts can extract energy from sugar either with the help of oxygen (respiratory metabolism) or without it (fermentative metabolism). Since only the latter produces alcohol, it is obviously in our interests to encourage fermentation over respiration. The more efficient (from the yeasts' point of view) respiratory metabolism takes place inside tiny sausage-shaped structures in the cell, the mitochondria. Although the vast majority of a eukaryotic cell's DNA is packaged in the cell nucleus, a small portion is found in mitochondria. This mitochondrial DNA contains the instructions needed to make some of the proteins involved in respiration. Some mutant strains of yeast lack mitochondria and are, thus, forced to rely on fermentation to meet their energy requirements. As this is less efficient, these mutants grow more slowly, forming only small colonies, and hence they are known as *petite* mutants. By comparing *petite* yeasts with normal yeast cells it was possible to garner a wealth of information about the way genes are arranged and how they operate. Such studies in the fifties and sixties helped pave the way for the new techniques of genetic engineering.

The relevance of this to fuel-alcohol production is that *petite* yeasts yield up to twice as much alcohol as their normal relatives. For example, one strain of ordinary yeast, coded IZ–1904, produces only 41 per cent of the theoretical maximum amount of alcohol from a given weight of glucose sugar, while the *petite* version of this yeast yields 83 per cent. Researchers in Brazil and elsewhere are already attempting to introduce *petite* yeasts into large-scale alcohol manufacturing processes.

## Alternatives to yeast

The production of alcoholic beverages in the West is based almost entirely on yeasts, but in other parts of the world several different microbes have been used for centuries to ferment sugar to alcohol. One such organism, the bacterium *Zymomonas mobilis*, has attracted particular attention. This bacterium has long been used in Central America to ferment the juice of the agave plant, producing a potent brew known as pulque. It appears to ferment sugar more efficiently than yeasts and its capabilities are now being investigated. The cylindrical cells of *Zymomonas mobilis* can be attached to fibres of cotton (Plate 12), and these immobilized cells offer many advantages over free-floating cells. In particular, systems that employ immobilized cells can be run for days, perhaps even weeks, without the need to stop the fermentation to add fresh cells. Such continuous fermentation processes are almost certain to have a great impact on fuel-alcohol production in the future as they will substantially reduce costs.

Both the US Department of Energy's Fuels from Biomass and the EEC's Energy from Biomass programmes are funding research into alternatives to yeasts. Two species of bacteria have aroused considerable interest: *Clostridium thermocellum* which feeds off cellulose; and *Thermoanaerobacter ethanolicus* which consumes starch and sugars. Their names provide a clue to their potential importance, for they are both thermophilic (heat-loving) bacteria. The overall costs of any industrial process are likely to decrease if the speed of throughput from raw materials to products is increased, and the rate of any chemical reaction increases as the temperature is raised. When dealing with complex and delicate living cells, there is no point in simply warming them in the hope that they will work faster. Yeasts have evolved to live in moderate temperatures and they are not spurred into action by extra heat; they just die. However, *Clostridium thermocellum* and *Thermoanaerobacter ethanolicus* have different life styles. The latter is particularly resistant to heat, and is normally to be found revelling in the near-boiling waters of hot springs in Yellowstone National Park in the US and Iceland.

The use of these heat-tolerant bacteria would bring four potential benefits for the biotechnological production of fuel alcohol. They should convert their raw materials (starch, sugars or cellulose) into alcohol more rapidly than yeasts do. There would be no need for complex cooling systems in the fermentation apparatus. (At present, fermentation vessels must be cooled to ensure that the heat produced during fermentation does not harm the yeasts.) Fermentation vessels operating at relatively high temperatures are less likely to become contaminated with unwanted organisms, since not many other microbes could stand the heat. Finally, as the fermentation liquid is already warm, less heat need be provided to distil the alcohol, and the cost of distillation is a significant factor in total production costs.

# Methane – fuel from wastes

Man's first acquaintance with inflammable gases was almost certainly the sight of shimmering lights low over marsh ground. These ethereal illuminations, called will o' the wisps, are created by marsh gas burning as it seeps out of the ground. Three thousand years ago, the ancient Chinese book, the *I Ching*, referred to fire in the marsh and it is believed that the people of the Sichuan region were using marsh gas as a fuel at this time. The major component of marsh gas is methane and, in recent years, it has come to play a very important role in our daily lives, notably as the main constituent of natural gas extracted from beneath the North Sea and many other parts of the world. Over millions of years the activities of many types of microbe have decomposed the dead cells of plants and animals, producing these vast reserves of methane.

Our dependence on microbes that produce methane has a more direct and surprising form – without them our supplies of meat would be drastically reduced. The chief food of cattle is grass, the cells of which are surrounded by tough walls of cellulose. Neither humans nor cattle can digest cellulose, but cattle can thrive on a diet of grass because their stomachs are populated by many types of bacteria. Between them these are able to break down the cellulose and produce methane, releasing the contents of the plant cell which the cattle can digest.

Now methane-producing bacteria (termed methanogens) are being scrutinized by biotechnologists to see how they can supply us with methane fuel. The general aim of these studies, as summed up by some Californian researchers, is to develop 'the fermentative production of methane from plant materials in the absence of the currently required fermentation apparatus (a cow)'.

As mentioned earlier, one of the keys to commercial success for a biotechnology is its use of readily available and cheap raw materials. The basic raw materials for methane manufacture are often all too abundant – domestic and farmyard sewage, and industrial effluents.

In all the biotechnologies discussed so far each process has used only a very few types of microbe, and often only one species. By contrast, methane-generating systems normally consist of many different types of microbe. This is not surprising, since the raw materials fed into the process are usually complex mixtures of materials, and it is unlikely that any one species could deal with all the different compounds. Thus, efficient methane generators involve a complex interaction between many types of microbe, each playing an important role.

The simpler installations, which are often little more than a pit in the ground, use mixed raw materials such as domestic and agricultural wastes. These are usually called digesters and their product is known as biogas. This rather vague term is used because the composition of the gas varies, although methane always forms the bulk.

The variability of the raw materials and products, and the diversity of microbes employed, poses problems for biotechnologists. Methane-generating systems are known to work, but it is not clear just how. To be able to design improved processes, biotechnologists would like to find out precisely which organism is performing each chemical reaction and how they all interrelate. Faced with a seething mass of ill-defined composition, this task is far from simple. Despite these difficulties, much information is now being accumulated which may aid in the design of better methane generators.

In the meantime, 'non-scientific' digesters are being installed in vast numbers throughout the world. The People's Republic of China is thought to have about 5 million simple installations in rural areas and the Indian Government's ambitious Gobar plan (from the Hindi for cow manure) has already brought digesters into tens of thousands of villages, providing a useful supply of cheap energy. This type of biotechnology is probably best suited to small-scale operations in which the raw materials are collected from the locality and the resulting power is distributed within the same area. In this way high transport costs – which could become prohibitive for large methane-generating systems – can be avoided. As such digesters are very simple constructions, most of the difficulties associated with high-technology industries are avoided, allowing a new source of energy to be introduced with the minimum of cost and difficulty.

More sophisticated, experimental methane-producing plants are in operation, particularly in the US. Most of these are based on two-stage processes that involve both algae and bacteria. The raw materials (again mainly sewage) are fed into shallow open ponds in which the algae grow. The algae are then harvested from time to time and introduced into a digester, where various bacteria consume the algae and produce methane. These two-stage systems have the special advantage that the growing algae use the sunlight for photosynthesis and, thus, more biomass is obtained than if the organisms were simply consuming the raw materials supplied. One of the main drawbacks is that large quantities of water are required and, in the places where sunlight is plentiful, water may be scarce. Fortunately, some algae thrive in brackish water, which is often in better supply. One such is the filamentous alga *Spirulina*, whose thread-like cells clump together, making it easier to harvest than most other microbes.

It appears that almost any type of plant material can be fed into methane digesters, and this fact has prompted an examination of many different schemes. Some seem rather fanciful, but might one day become practicable. For instance, there has been a proposal to grow giant kelp (a type of seaweed) on vast floating grids in the oceans. The rate of growth of kelp can be quite dramatic – up to 14 per cent a day. Obviously the practical problems include damage by storms, harvesting the kelp, and producing the methane at sea or transporting the kelp to land. It is unlikely that this particular scheme will prove feasible, but it provides a good example of the ambitious nature of projects being investigated in this area of biotechnology.

## Hydrogen – the perfect renewable fuel?

The excitement aroused by biotechnological research into hydrogen production is not surprising, for it offers a fuel that can be produced from a power supply that will last for millions of years (the sun) and a raw material that covers three-fifths of the globe (water). In addition, hydrogen causes no pollution when burnt, but forms water, thus renewing the raw material.

The public's view of hydrogen is still coloured by the memory of the explosion of the *Hindenburg* airship. In fact, hydrogen is scarcely more dangerous than many fuels used in the home. It is mainly cost rather than safety that has prevented hydrogen from becoming a common fuel. If the efforts of biotechnologists bear fruit, this could change in the next few years.

Nearly all the materials inside a cell contain hydrogen atoms, and a great many of the cell's vital chemical reactions involve the transfer of hydrogen from one compound to another. Despite hydrogen's ubiquity, only a few organisms release hydrogen gas into their surroundings. Gaseous hydrogren consists of two linked hydrogen atoms ($H_2$). Most of the cell's hydrogen is joined with other atoms to form many different compounds, or exists as hydrogren ions ($H^+$, hydrogen atoms that lack electrons). Some microbes possess enzymes that can take two hydrogen ions and two electrons, and then join them to form a molecule of hydrogen gas, and these microbes have captured the attention of many biotechnologists.

The enzyme responsible for this intriguing method of making hydrogen gas is called hydrogenase, and so far it has been extracted from about fifteen different species of bacteria and some algae. Attempts to put these organisms to commercial use are still at a very early stage, but there is considerable optimism about the prospects for success. Laboratory-scale experiments show that hydrogen is produced when these bacteria, for instance, *Clostridium butyricum*, are supplied with sugars. These systems tend to be unstable, and after a while the bacteria stop making hydrogen. No-one knows precisely why the process falters, although, presumably, the conditions under which the bacteria are kept do not match those of their natural habitat. Further research may overcome this difficulty, and encouraging indications are to be found in the work of some Japanese biotechnologists who have immobilized cells of *Clostridium butyricum* on filters and fed them with waste water containing sugars from an alcohol factory. The bacteria continued to produce hydrogen gas for over a month, compared with only a few hours in the earliest experiments.

An alternative approach is to obtain hydrogen gas from water by an ingenious combination of three distinct biotechnological techniques. Such systems have already begun to prove their worth in laboratory experiments. When illuminated, the alga *Chlorella pyrenoidosa*, can assemble two water molecules and two carbon dioxide molecules to form a compound known as glycolate. This substance, collected by breaking apart the algal cells, is then used as the raw material for the next stage. Glycolate is fed to an immobilized enzyme, glycolic oxidase, obtained from plants. This enzyme catalyses the transformation of glycolate into formate. In the final stage of the process, formate is supplied to bacteria which have been fixed on to glass beads, and the bacteria consume the formate, releasing hydrogen gas and carbon dioxide. Thus, the sum total of all three stages has been to separate the hydrogen and oxygen atoms that were originally combined in water molecules (see Figure 1). This rather devious approach to an apparently simple process is necessary because it is not practicable to break apart water molecules in a single step.

Biotechnologists are taking a keen interest in another method of making hydrogen, employing the photosynthetic machinery of plants. Without photosynthesis, virtually all life on Earth would eventually cease. The complex interlocking food chains on which all animals depend can be

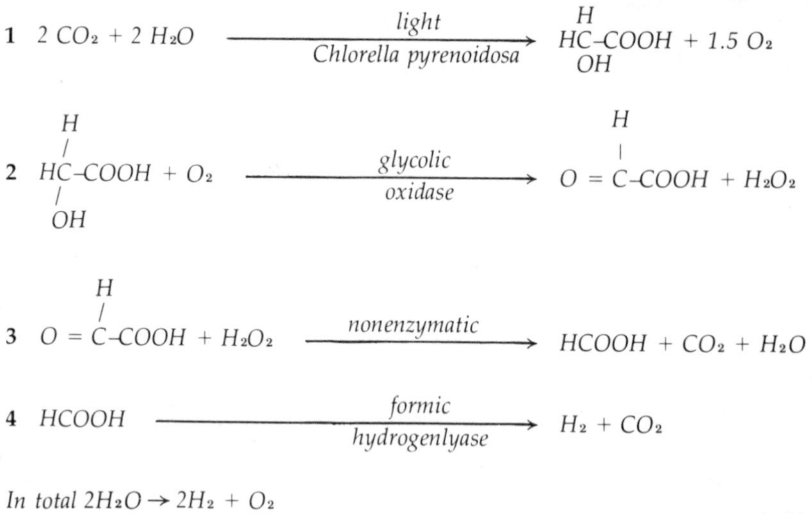

$$1 \quad 2\,CO_2 + 2\,H_2O \quad \xrightarrow[\text{\textit{Chlorella pyrenoidosa}}]{\text{\textit{light}}} \quad \underset{OH}{\overset{H}{HC}}\text{-}COOH + 1.5\,O_2$$

$$2 \quad \underset{OH}{\overset{H}{HC}}\text{-}COOH + O_2 \quad \xrightarrow[\text{\textit{oxidase}}]{\text{\textit{glycolic}}} \quad \overset{H}{O = C}\text{-}COOH + H_2O_2$$

$$3 \quad \overset{H}{O = C}\text{-}COOH + H_2O_2 \quad \xrightarrow{\text{\textit{nonenzymatic}}} \quad HCOOH + CO_2 + H_2O$$

$$4 \quad HCOOH \quad \xrightarrow[\text{\textit{hydrogenlyase}}]{\text{\textit{formic}}} \quad H_2 + CO_2$$

*In total* $2H_2O \rightarrow 2H_2 + O_2$

**Figure 1**

A three-stage method for making hydrogen gas from water

traced back to photosynthetic organisms.* Plants, algae and some bacteria are the only organisms that can tap the sun's power to provide the energy necessary to maintain life.

As befits such a vital process, the intricacies of photosynthesis have been studied in great detail by biochemists and plant physiologists, and its outlines are now well-understood. When certain molecules (notably chlorophyll) are illuminated, they absorb energy. This energy is transferred from one part of the cell to another and is ultimately converted into a form the cell can use, mostly ATP. Once the energy has been harvested from sunlight, it is used to combine water and carbon dioxide (from the air) forming the complex organic molecules the cell needs. In the first part of this process, water molecules are split and oxygen is released into the atmosphere.

A host of different molecules are involved in each of the many stages of photosynthesis, and by no means all the interrelationships between these substances are known. However, we are chiefly concerned with just one aspect of photosynthesis – the process by which water molecules are split. Plants do not make hydrogen gas ($H_2$); instead water is decomposed to yield oxygen atoms, electrons and hydrogen ions ($H^+$). In a normal plant cell the $H^+$ ions do not get a chance to combine to form $H_2$, but are used to manufacture energy-rich compounds. To produce $H_2$ this process must be subverted. This presents a very considerable challenge since evolution has ensured that the different steps in photosynthesis are subtly interconnected, so promoting or preventing particular processes is a formidable task. Despite this, biotechnologists in several countries have produced some very promising designs.

Most of these experimental systems are based neither on whole plants, nor even on entire plant cells. Instead they use tiny oval structures, known as chloroplasts, extracted from plant cells. These structures contain chlorophyll and a variety of other compounds that capture and channel light energy. Figure 2 shows the basic processes that take place during the production of hydrogen in a typical experimental system. So far, systems like this have only been made to work for a few hours. However, only about six years ago the systems were faltering after only minutes. If this rate of progress can be maintained, a commercially feasible process will come into operation in the future. Investigations are underway to find the most resilient component for each stage of the process. For example, the chloroplasts often used are obtained from spinach, but certain weeds seem to have more stable chloroplasts and may thus prove more suitable. A

---

* It is now known that some microbes can extract energy from inorganic compounds found in rocks and seawater (see page 164). These probably do not depend on the activities of photosynthetic organisms but they constitute only a tiny proportion of the planet's life forms.

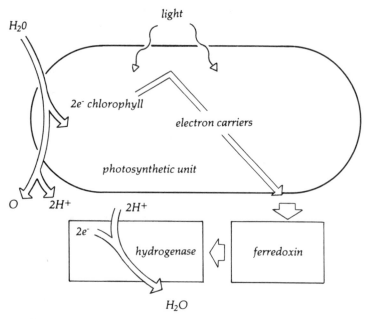

**Figure 2**

Chloroplasts extracted from plants capture light energy with the aid of chlorophyll molecules. The energy is used to split a water molecule into two $H^+$ ions and an atom of oxygen, and simultaneously two electrons ($e^-$) are released. The $H^+$ ions diffuse from the chloroplasts to hydrogenase, and the electrons are transported by a series of molecules known as electron carriers. In experimental hydrogen-producing systems, hydrogenase enzymes, obtained from bacteria, are used to form an $H_2$ molecule from two $H^+$ ions and two electrons. The hydrogen gas so formed, bubbles out of the solution and is collected.

major problem with the hydrogenase enzymes is that many are very sensitive to oxygen, and oxygen is produced when water is split. Hydrogenases from certain bacteria have been found to be more robust and are not disabled so rapidly by oxygen.

It may turn out, when all the options have been examined, that purely biological materials are just too fragile for this industrial application. This would not mean that years of research had been wasted for a more detailed understanding of how nature has solved certain problems provides valuable clues about how we may do the same. For instance, impressive advances in the construction of artificial photosynthetic systems modelled on natural ones are now being made. Biomimicry – learning from and imitating nature – is set to become a major facet of technology in the years to come.

# CHAPTER EIGHT

# Biotransformations
# – the way ahead for industry

Shell, Du Pont, ICI, Exxon, International Nickel and Standard Oil are just a few of the multinational conglomerates that have been tempted to take a stake in the future of biotechnology. Many have interests in the pharmaceutical, agricultural and energy industries and, therefore, stand to benefit from the biotechnological developments already described, but they must be particularly attracted by the potential of biotechnology in their other business activities – the chemical, mining and oil extraction industries which lie at the core of the developed world's economy.

At present much of the world's chemical industry relies on raw materials obtained from oil, gas and coal. Plastics, paints, adhesives and synthetic rubbers are all derived entirely or in part from these diminishing resources. In 1978, in the US alone, over $35 billion worth of petrochemicals were made from relatively few raw materials. Several of these can already be manufactured by biotechnological processes, including acetone, glycerol, butanol and ethanol, and the rapid progress in this field may soon yield others.

## Acetone and butanol – a biotechnological renaissance?

War is unfortunately one of the most potent catalysts for industrial innovation. Soon after the outbreak of the First World War, Britain and Germany faced a similar crisis: chemicals required for the manufacture of munitions were in short supply. Britain needed more acetone, a liquid solvent used during the production of cordite, while Germany lacked glycerol, from which dynamite is made. Both countries turned to microbes to solve their problems and met with considerable success. Germany was soon making 1000 tonnes of glycerol a month. After the war, the rise of the petrochemical industry supplanted biotechnology as the major source of acetone and butanol. In recent years, a better understanding of the microbes that make these chemicals has been gained and the cost of oil has increased substantially. These factors may bring us full circle, reinstating these and other biotechnologies in a central position in the chemical industry. This time they can be put to much more constructive uses, including the manufacture of plastics, fibres and resins.

In 1912, the chemist Chaim Weizmann (who was later to become Israel's first president) was working in Manchester. He developed a process by which the bacterium *Clostridium acetobutylicum* fermented starch to form the liquids acetone and butanol, and it was this process that was introduced on a large scale during the war. While Weizmann's process has largely fallen into disuse, the demand for acetone and butanol has grown unabated. Over a million tonnes of acetone are consumed by the US chemical industry each year, much of it in the manufacture of plastics. Butanol, a kind of alcohol, is even more versatile; it is employed in the manufacture of resins, protective coatings, paints, synthetic rubber and brake fluids.

In many respects the Weizmann process is similar to the production of fuel alcohol (see page 144), and biotechnologists face the same kind of problems in developing economically attractive systems. There is one particularly significant difference: when yeasts are supplied with sugars in the absence of air, they will always make alcohol, but *Clostridium acetobutylicum* does not always make acetone and butanol. Microbiologists are searching for strains of the bacterium that can be relied upon to produce the desired products. However, this bacterium is unusually sensitive to the conditions under which it is grown and unless the temperature, acidity and other factors are precisely right it will not yield acetone and butanol in significant quantities. This means that the fermentation must be very carefully controlled.

Another difficulty is that the bacteria are harmed by the products they make. Just as yeasts cannot tolerate high concentrations of alcohol in their surroundings, so *Clostridium acetobutylicum* is damaged by acetone and butanol. Unfortunately, here the problem is even more acute, and only 2 or 3 per cent of these materials in the fermentation liquids will inhibit the bacteria from making more of them. This means that large fermentation vessels must be built to obtain the desired quantities of products, and the cost of purifying such a small proportion of acetone and butanol from all the other materials in the vessel is high.

Despite these obstacles, there are good grounds for optimism about the future of the Weizmann process. In particular, recent experiments indicate that immobilizing the bacteria may increase the efficiency of the fermentation by 65 to 200 per cent. If this increase is achieved in practice, the economics of the process will improve radically.

Until the First World War, German industry had manufactured glycerol from imported vegetable oil, but the British naval blockade soon began to stem the flow of the raw materials. This provided the impetus to develop the work of the biochemist Carl Neuberg who, a few years before, had discovered that the small amount of glycerol produced by yeasts when fermenting sugars could be greatly increased by the addition of a simple, cheap chemical – sodium bisulphite. Glycerol has since grown to become an almost indispensable part of the chemical industry, being employed, for example, as a lubricant and softener, a plasticizer for cellophane and a raw material for resin manufacture. Today, Neuberg's process has sunk into commercial obscurity and glycerol is made from petrochemicals and vegetable oils. In a few years, his process may be revived, again due to the rising costs of alternative manufacturing methods. Rather nearer at hand is an entirely novel biotechnological source of glycerol – algae.

Despite its forbidding name, the Dead Sea has been colonized by a few unusually hardy microbes. The high concentrations of salt (up to eight times that of the oceans) in the Dead Sea and the Great Salt Lake of Utah soon kill most organisms. They die as the water inside their cells is pulled out through the membranes into the highly saline environment. Certain microbes, called halophiles (salt-lovers), have evolved ways of preventing this dehydration. One of these, the alga *Dunaliella bardawil*, manufactures large quantities of glycerol so that the cell contains a high concentration of dissolved material. This counteracts the force – known as osmotic pressure – which tends to draw water out of the cells and into the surroundings.

In Israel, near the Red Sea coast, this alga is grown in specially constructed ponds covering 2 hectare (5 acres). Since *Dunaliella bardawil* is a photosynthetic organism, it obtains most of its energy from the sun, and only a few simple nutrients need to be supplied. Once the algae have been harvested and dried, the glycerol, which accounts for something approaching 40 per cent of the weight of the cells without their water is extracted.

This alga is one of the most promising organisms for biotechnology. It also contains about 8 per cent of beta-carotene, the compound which gives carrots their characteristic colour and finds a ready market as a food colourant. Once the glycerol and beta-carotene have been taken out of the cells, the residue forms an excellent, protein-rich animal feed. In addition, this alga thrives in brackish water, and so it can be grown in semi-arid regions where fresh water is at a premium. The fact that this type of biotechnology does not compete with agriculture for good-quality water is a particularly important factor for many less-developed countries. *Dunaliella bardawil* lends itself to low-technology industries which need little capital investment. Valuable products can be obtained from the cells fairly easily and the growth ponds do not require much attention. In particular, contamination by unwanted organisms – the bane of most biotechnologists – is largely eliminated. Most other organisms are destroyed by the very high concentrations of salt found in the ponds.

# Biotechnology and the plastics industry

The plastics industry is currently worth over $50 billion a year – a tempting market for biotechnological enterprises. As we have seen these could supply acetone, glycerol and butanol to the chemical industry, and the ethanol produced by fuel alcohol factories could also be diverted for use in it. Ethanol serves as a vital starting material for the manufacture of matierals as diverse as detergents, dyes, adhesives and resins for synthetic fibres.

Among several novel and ambitious projects, one of the most promising is the synthesis of compounds known as alkene oxides which are widely used in the manufacture of plastics and polyurethane foams commonly used in furnishings. Development of these processes began about five years ago and may be in commercial operation before the end of this decade, with potential sales of $2–3 billion.

Alkenes are a group of compounds which contain only carbon and hydrogen. Most importantly for the plastics industry, alkenes can be polymerized – that is, the individual molecules can be linked in chains – to form materials such as polypropylene (used, for example, to make containers) and polyethylene (better known as Polythene). Before alkenes are polymerized to form plastics, they must be converted into alkene oxides by the addition of oxygen to the molecules. At present, this is accomplished by purely chemical techniques, but two biotechnological processes are on the horizon. The first, largely developed by the Californian firm Cetus, employs three enzymes – two from fungi and one from bacteria – to perform the same function as the current chemical methods of making alkene oxides (see Figure 1).

More promising still is a process patented in 1981 by scientists from the University of Warwick, England. In the famous baths of the city of Bath, they discovered a microbe which can add oxygen to alkenes. When supplied with propylene or ethylene gases, the bacterium *Methylococcus capsulatus* can insert an oxygen atom into the molecules to produce propylene oxide or ethylene oxide. This bacterium is particularly enticing because it lives very happily at about 45°C (113°F), and at this temperature the alkene oxides are gaseous. It is much simpler to collect the product as a gas than as a liquid, mixed in with all the other materials in a fermentation vessel.

Both the enzyme and the bacterial systems have several advantages over conventional processes. The current chemical methods require expensive chlorine gas, whereas the enzyme technique employs cheap common salt and the bacteria do not even need that. Both biotechnological processes work at lower temperatures, saving on energy costs, and they are flexible in that different types of alkene oxides can be made with the same basic installation. Finally, the pollution created by chemical treatments can be avoided by employing enzymes or bacteria.

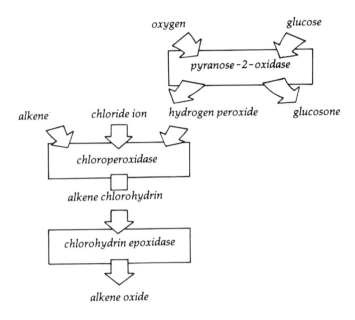

**Figure 1**
Alkenes, such as ethylene and propylene, are converted into alkene oxides
and then polymerized to form plastics, such as polyethylene and
polypropylene. A series of three enzymes can be employed to make alkene
oxides. First an enzyme, pyranose-2-oxidase (obtained from fungi)
catalyses a reaction between oxygen and glucose to produce hydrogen
peroxide. Another fungal enzyme, chloroperoxidase, joins hydrogen
peroxide and a chloride ion from sodium chloride (common salt) to an
alkene chlorohydrin molecule. The chloride and a hydrogen ion are then
removed with the aid of an epoxidase enzyme from bacteria to yield an
alkene oxide.

A notable aspect of these future biotechnologies is the way that they
would be integrated into conventional industries. The alkenes would still
be derived from oil or similar materials, and the polymerization of alkene
oxides to produce plastics would still be achieved by chemical techniques,
with biotechnology making its contribution at the intermediate stage. The
reliance of the process on oil as the source of alkenes is an obvious
drawback. Looking much further into the future one can envisage
biotechnology supplying the alkenes. No known microbe makes
significant amounts of alkenes, but, as has happened so often before, a
diligent search for organisms that contain the required substances may
prove successful. If such a search failed, the problem could be turned over

to genetic engineers, for whom designing a microbe to supply alkenes would be a major challenge. There is very little chance that the introduction of a single gene into a microbe could induce it to start synthesizing alkenes, since genes code for enzymes and other proteins, compounds that are quite different from alkenes. It would be necessary to put several genes into a microbe, each of which codes for one of a series of enzymes. These enzymes might then act in concert to convert some substance normally found in the cell into alkenes.

A rather more immediate biotechnological process for the plastics industry involves a compound called polyhydroxybutyrate (PHB). This material, which is similar to synthetic polyesters used in the textile industry, is found in many types of bacteria. Bacterial PHB is rather fragile, but it is biodegradable. This has stimulated thoughts of using PHB in surgical sutures; threads of this material inserted during operations would later dissolve once their job was done. ICI plans to scale up its production of PHB in the near future.

## Sweetening the profits from oil wells

In the market for fuels and raw materials for the chemical industry, biotechnology and the oil industry will be in fierce competition. In sharp contrast to this, biotechnology promises to help oil drillers recover vast oil reserves. When a driller first 'strikes it rich' the oil gushes to the surface, but, unfortunately, less than half the contents of a typical oil field comes to the wellhead with such ease. Extracting the rest requires considerable skill and ingenuity. About 50 per cent of the oil now being produced in the US is taken out of the ground with the aid of so-called secondary recovery methods. These may involve, for example, pumping water into the rock formation forcing the oil along subterranean crevices until it reaches the collection wells. Even using secondary recovery £300 billion worth of North Sea oil and 90 per cent of the US's 330-billion-barrel oil reserves may remain stubbornly locked underground. Tertiary oil recovery methods are needed to extract this precious resource, and this is where biotechnology steps into the picture.

Most of the oil in a reservoir is found not as vast pools of liquid, but as a coating on grains of rock. The oil sticks tenaciously to these grains and must be dislodged before it can be brought to the surface. Ordinary water is too thin a liquid to budge most of this oil; it simply flows past the oil-coated grains. In tertiary oil recovery (also known as enchanced oil recovery) materials are mixed with the water to make it more viscous, and one such material is xanthan gum.

Xanthan gum, a polysaccharide produced by the bacterium *Xanthomonas campestris*, is composed of many glucose units linked into a chain with other types of sugars branching off at regular intervals. This knotty structure

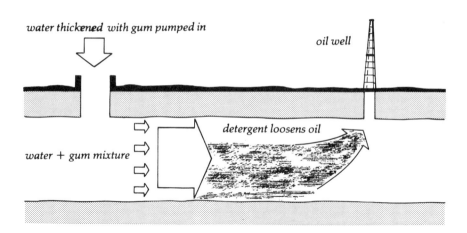

**Figure 2**
Using xanthan gum to recover oil. Water containing a detergent-like
material is first pumped into the ground to loosen the oil clinging to rock
particles. Water thickened with xanthan gum is then pumped in to act as a
piston, pushing the oil-bearing mixture towards the oil well.

makes xanthan gum a very efficient thickening agent, allowing the
water/gum mixture to act like a piston, pushing oil towards the wells (see
Figure 2). Price is the major obstacle at present to its general use in
extracting recalcitrant oil deposits. The kind of biotechnological strategies
that have been applied so successfully with other organisms could easily
reduce the cost of xanthan gum quite dramatically in the near future.

Xanthan gum and other similar materials excreted by microbes have
already found a use in the petroleum industry. The huge drills which
penetrate rock must be lubricated, and various types of mud perform this
task. These drilling muds – mixtures of water, clays and other materials –
also counterbalance the upward pressure of the oil. Microbial
polysaccharides have been proved to be eminently suitable thickening
agents.

If xanthan gum is likely to be so effective in tertiary oil recovery, why go
to the trouble of growing the bacteria in factories, extracting the gum and
then pumping it into the oil fields? Why not send the microbes themselves
underground? This is exactly the line of thought being pursued by some oil
companies. Field tests employing *Bacillus* and *Clostridium* bacteria rather
than *Xanthomonas* are already under way, and the preliminary results are
encouraging. Sugars and other nutrients are fed to these bacteria while
they are deep below ground. Here they grow, produce chemicals that help

wash the oil free, and yield carbon dioxide and other gases which also assist in forcing the oil to the surface.

It is remarkable that these bacteria can survive in conditions of high pressure and temperature, lack of water and oxygen, and large quantities of salt and sulphur, all of which are inimical to these organisms. This shows again that the resilience and adaptability of microbes should never be underestimated, and these qualities, enhanced as necessary by biotechnologists, may soon be put to commercial use in the oil industry.

## Microbes in mineral mines

The ore extraction plant of the future could have the appearance of a present-day water-treatment plant . . . free from the dirt and spoil heaps normally associated with mining operations, while far below ground millions of microbes are carrying out the tasks which today are characterized by the roar of machinery and the ring of pick and shovel on rock (Dr Richard Manchee, Microbiological Research Establishment, UK, 1979.)

This view of mineral mines is not a fantasy; microbes are already being used to extract valuable metals, such as copper and uranium, from rocks. Over the next few years it is almost certain that microbial workforces will be employed in many more mining operations.

In conventional mining for metallic minerals, vast quantities of rock must be crushed and ground, then treated with chemicals to recover the metals. Although laborious and expensive, this time-honoured process is economically worthwhile *if* the rocks contain high-grade ores – that is, if they contain a relatively high proportion of the desired metal. High-grade ores are increasingly in short supply, but there are still large amounts of metals in the ground. For example, the gold mines of Wales and California have been generally abandoned, not because all the gold has been removed but because the rich seams have been exhausted and it is not worth the effort to process ever larger amounts of rock to obtain the same yield of precious metal.

Thus, the problem that faces the mining industry is to make use of low-grade ores which cannot be exploited economically by conventional techniques. Mining engineers in Spain, Canada, the US and elsewhere have found that the answer is to rely on microbes to extract the valuable metals and concentrate them into a form that is cheaper and more convenient to handle.

*Thiobacillus ferro-oxidans* is probably one of the oldest forms of life on Earth, but it was not until 1947 that it was discovered in an abandoned coal mine in West Virginia. This bacterium is now known to be present in many types of rock throughout the world, and there can be many millions in just a handful of material. Many microbes have bizarre dietary preferences, but none so odd as this rod-shaped organism. It does not obtain energy from sunlight (it usually lives in total darkness) nor from organic materials in its surroundings. Instead, it unlocks energy from inorganic compounds, such as iron sulphide, and uses that energy to construct the materials it needs to live from the carbon dioxide and nitrogen in its environment. In the process it also manufactures sulphuric acid and iron sulphate, which explains why *Thiobacillus ferro-oxidans* can be used in mining operations.

The sulphuric acid and iron sulphate it produces attack the surrounding rocks and leach (dissolve) many metallic minerals. For instance, the activities of these microbes will convert insoluble copper sulphide into soluble copper sulphate. As water percolates through the rocks the copper sulphate is carried along and eventually collects as bright blue pools. In this way the copper scattered throughout thousands of tonnes of low-grade ore is concentrated in metal-rich lagoons. The metal is recovered by passing copper sulphate solution over pieces of iron. Eventually a layer of copper is deposited on the iron, and this can be scraped off. Uranium is leached from its ores by the same type of process.

Already about 14 per cent of the copper produced in the US depends on this biotechnology. At present, microbial leaching is mainly employed with the waste materials from conventional mining and extraction processes which leave substantial residues of metal in the discarded rocks. Dumps, up to 370m (1200ft) high and weighing 4 billion tonnes, are constructed from these 'waste' materials (see Plate 13). Water is sprayed on to the top of the mounds and, as it filters down, it picks up the soluble metal compounds created by the action of the bacteria. The ubiquity of *Thiobacillus ferro-oxidans* means that it is rarely necessary to introduce it into the dumps. The area on which the heaps are built is usually covered with clay or asphalt so that the metal-rich liquids collect in pools at the foot of the heaps instead of seeping right into the ground.

When using microbes in this type of secondary recovery process, miners are still faced with the substantial cost of hauling the ores to the surface. The experience gained at the Stanrock Uranium Mine in Canada shows that even this cost is not always necessary. This mine was opened in 1958, operating on conventional principles. By 1962 it was found that the pools of liquid that had accumulated underground contained about 13,000kg (29,000lb) of uranium oxide which had been leached out of the rocks. Before long, conventional mining was halted and bacteria were left to do most of the work, with water being hosed on to the rocks to assist the natural leaching process. This underground solution mining has cut costs by a quarter at this mine. Similar techniques are almost certain to be introduced into other mines, especially those with low-grade ores. Semi-industrial processes have already shown the promise of microbial leaching in the recovery of cobalt, lead and nickel, and other valuable metals, such as cadmium, gallium, mercury and antimony, are targets for the future. The potential is truly enormous – for example, the US has several billion tonnes of rock which contain small concentrations of nickel. Since there is, on average, only about 1kg (2.2lb) of nickel in each tonne of rock, it is uneconomic to extract this valuable resource with conventional mining techniques. Advances in microbial mining could make it possible to tap this source of nickel which is worth about $60 billion and recover a further $10 billion hoard of cobalt in the same rocks.

Microbial mining has the additional advantage for less-developed countries that it eliminates the need for some of the costly (and usually imported) heavy mining equipment. Although a substantial capital investment is required for any mining development, microbes could help these countries preserve precious foreign currency.

## Microbes clean up the mess

The threat that pollution poses to ourselves and the environment has become all too apparent in recent decades. Oil spills, pesticides, herbicides, chemical effluents and heavy metals, such as lead and mercury, are just some of the hazards. Biotechnology can be used to tackle these problems in two ways. Firstly, the root causes can be attacked by the introduction of more biotechnological production methods, which are intrinsically less polluting. For example, in manufacturing chemicals for the plastics industry with the aid of biotechnology, microbes are fed on innocuous raw materials such as sugar, whereas conventional processes use oil-based raw materials, some of which inevitably escape to pollute the environment. Secondly, microbes can be deployed as voracious scavengers, removing all manner of pollutants.

The principal aim of all the biotechnologies discussed so far is to manufacture specific products; the fact that certain raw materials are incidentally consumed during the process is an often expensive necessity. In biotechnological processes that are targeted towards the control of pollution, the emphasis is reversed; their primary purpose is the destruction of specific raw materials – the pollutants. Despite this fundamental distinction, it should be noted that there is no precise dividing line between product-orientated and raw-material-orientated biotechnologies. For instance, the economic feasibility of the methane generators described in Chapter 7 depends on two factors: the value of the methane fuel produced, and the fact that they perform a useful function in disposing of domestic and agricultural wastes.

### Pouring microbes on troubled oil

The *Pseudomonas* are a group of bacteria noted for their ability to break down esoteric compounds that most microbes shun. In particular, various strains of *Pseudomonas* can consume hydrocarbons, which constitute the bulk of oil and petrol. However, each individual strain can utilize only one or a few of the many different types of hydrocarbon. The genes that code for the enzymes which attack hydrocarbons are not found on the main bacterial chromosome, but on plasmids, the small, semi-autonomous rings of DNA.

In an attempt to create a 'superbug' which would be able to mop up *all* the types of hydrocarbon in oil spills, Ananda Chakrabarty of General Electric introduced plasmids from several different strains of *Pseudomonas* into a single cell (see Figure 3). One idea was to grow these recombinant bacteria in the laboratory, mix them with straw and dry them. The bacteria-laden straw could be stored until needed, when it would be scattered over oil slicks; the straw would first soak up the oil, and the bacteria would break it down into harmless, non-polluting materials.

Chakrabarty's microbe has not yet been used commercially, and inded many scientists doubt that it has much of a future; they argue that a mixture of wild strains of *Pseudomonas* may degrade oil just as well or even better. Whatever the merits of this particular bacterium, the general approach should prove valuable for controlling other types of pollution. Late in 1981, Chakrabarty and his colleagues announced that they had developed a microbe that attacked 2,4,5,-T, a very persistent herbicide and the main ingredient of the Agent Orange, which was used to destroy vast areas of jungle in Vietnam.

Although the suitability of genetically engineered microbes has yet to be proved, more conventional types of biotechnology have already been put to good use, as two examples from the US illustrate. When the *Queen Mary* was moved to Long Beach in California about 3,600,000 litres (800,000gal) of oily water lay in its bilges. Obviously, if this had been discharged into the harbour it would have harmed marine life and disfigured nearby beaches. Therefore, a mixture of several different sorts of bacteria was introduced into the bilges. In six weeks they decomposed the oil, leaving a combination of water, bacteria and innocuous chemicals that could be released safely into the harbour.

An oil company in Pennsylvania faced a similar problem when a leakage of 27,000 litres (6,000 gal) of petrol threatened to contaminate underground water supplies. Bacteria already living in the vicinity would doubtless have destroyed the petrol eventually, but without human intervention the process might have taken decades. The bacteria could only grow slowly because there were insufficient nutrients in their surroundings to give them the oxygen, nitrogen and phosphorus required for rapid growth. By pumping the missing nutrients into the ground, the bacteria were spurred into action and the petrol was degraded in only a year.

The living world has developed complex and highly efficient means of consuming natural wastes and utilizing them. If it were not for the constant recycling of materials – in which microbes play the major role – we would all be knee-deep in dinosaur remains! It is the new types of pollution, created by industry, that present the most intractable problems. New chemicals, such as pesticides and substances which previously appeared on the Earth's surface only in small amounts, such as oil and many metals, tend to persist since few of the common microbes in soil or water can use them as food. This has created a growing demand for tailor-made packages of microbes which can remove specific forms of pollution. Several major companies, especially in the US, offer mixtures of microbes and enzymes designed to clean up chemical wastes, including oil, detergents, waste waters from paper mills, and highly toxic materials such as dioxin, the chemical that wreaked havoc on the town of Seveso, Italy. By 1978, the US market for biological pollution control products had reached about $3 million. As public disquiet about the effects of pollution begins to be translated into more effective legislation, the demand for these products seems sure to grow – perhaps to fifty times its present level.

## Controlling pollution at source

The deployment of biotechnology to minimize the effects of oil spills and other pollution emergencies is clearly very valuable, but a far sounder approach is to attack the sources of pollution. Just about every city and town in the developed world has already made a substantial investment in one form of biotechnology – sewage processing plants. The majority of these depend on the action of microbes to purify waste water by consuming a wide variety of solid materials in domestic and agricultural effluents. Perhaps surprisingly, there is still no clear understanding of precisely how sewage is broken down. This is because the composition of sewage varies considerably and there are many different species of microbe at work in sewage ponds. However, as the existing techniques are usually very successful, there is little incentive to investigate new methods of disposing of general sewage. The situation is quite different in the case of certain industrial pollutants, most notably the heavy metals.

Heavy metals, a group of elements which includes mercury, cadmium and lead, are among the most insidious pollutants produced by modern industry. Mercury, for example, was responsible for the most notorious outbreak of metal poisoning, in which dozens of people in the Japanese fishing village of Minamata died or suffered severe damage to their nervous systems. Mercury discharged from a nearby factory was taken up by fish, which were later eaten by the local people. The dangers of lead pollution, particularly mental retardation of children, are beyond dispute. Many countries, Britain being a notable exception, have already introduced strict controls on the emission of lead from petrol engines, but there is also a need to remove lead from certain factory effluents.

Heavy metals are as toxic to most types of microbe as they are to animals

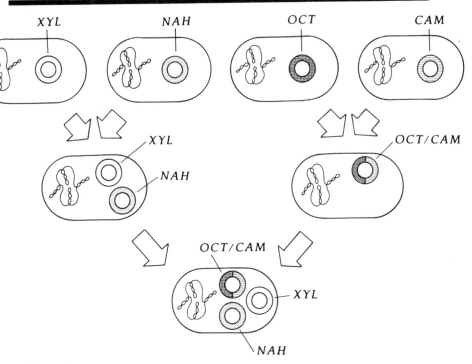

**Figure 3**
Oil contains several different types of hydrocarbon compounds, the main groups being xylenes, naphthalenes, octanes and camphors. Certain strains of *Pseudomonas putida* bacteria can consume each of these hydrocarbons, but no single strain found in nature can consume all four types. The genes which enable these bacteria to feed on hydrocarbons are found on four types of plasmids, termed *xyl*, *nah*, *oct* and *cam*. By introducing all four sets of genes into a single cell a 'superbug' was created which could digest all four major components of oil. The *cam* and *oct* plasmids cannot coexist inside the same cell, so the relevant genes from each plasmid are first joined into a single plasmid.

including humans, but some species of algae and bacteria avidly extract metals from their surroundings. This strange behaviour has prompted researchers in universities and industry to fashion new biological methods of purifying metal-laden effluent to replace the existing, and costly, chemical techniques. The idea is to grow bacteria or algae in ponds filled with factory effluents and suitable nutrients, allowing the microbes to scavenge metals from their environment and sequester them inside their cell membranes. A strain of the very common bacterium *Pseudomonas aeruginosa*, for example, is known to accumulate great quantities of uranium. As much as half of the total weight of each cell (excluding its water content) may be made up of uranium. These, and microbes which extract other heavy metals, could be filtered from their liquid environment and deposited in special dumps.

The same kind of approach could be used to recover valuable metals for re-use in industry. For example, *Thiobacillus* bacteria, similar to those that leach metals out of rock ores, can accumulate silver. A certain amount of silver inevitably appears in the waste waters from film factories and other industrial sites, and these bacteria could reduce the losses of this expensive resource.

In the last decade, acid rain has come to be recognized as a major threat to the environment. This pollution is caused when materials which contain sulphur, especially coal, are burned. The sulphur is oxidized and emitted into the atmosphere, where it dissolves in water droplets to form sulphuric acid. The acid rain accumulates in lakes, often hundreds of kilometres from the source of the pollution, killing fish and plants in vast numbers. Hundreds of lakes have been devastated by acid rain, especially in Canada and Scandinavia.

The spread of acid rain can be prevented either by removing sulphur from the coal before it is burned or by catching the sulphur oxides before they reach the atmosphere. Conventional chemical techniques for cutting sulphur pollution from coal add about £10 a tonne to its cost, but this might be halved if bacteria were called in to do the job. Several types of bacteria, especially those that inhabit hot-water springs, have a great appetite for sulphur compounds, from which they derive energy. Their activities would separate much of the sulphur from high-sulphur coal, producing a much cleaner and more valuable fuel. A small-scale plant in Ohio has shown that this process works, and in a few years' time the first industrial-scale operations may commence.

# CHAPTER NINE

# Who will benefit from biotechnology?

The ramifications of the bioindustrial revolution will extend far beyond the industries directly affected by it. The potential benefits – better health, more food, renewable energy sources, cheaper and more efficient industrial processes and reduced pollution – are immense, but what of the potential disadvantages? Any major new technology has profound social, economic and political effects. Biotechnology is no exception, and the possible consequences of the growth of biological industries on the health of workers and the public, on national and international trade, on economic power and on the position of science in society need to be examined.

## Genetic engineering – the sound and the fury

In the seventies, there erupted a controversy the vehemence of which has only been matched in science by the debate about nuclear power. The realization that humans had the power to transfer genetic material between completely unrelated organisms led to speculation about the possible creation of 'killer bugs'. The emotional atmosphere of the genetic-engineering debate was further heightened by an ignorance of biology among some who professed to represent the 'public interest' and a disturbing arrogance of some 'scientific experts'. In the mid-seventies, rational public debate was the exception rather than the rule.

In time, tempers cooled and, most importantly, more facts were accumulated, making it possible to produce reasonable assessments of the potential risks of specific types of genetic engineering – and how to reduce them. The central concern of those who wanted to regulate, or ban, genetic engineering research was that newly created organisms might produce uncontainable diseases. Two aspects of genetic engineering were claimed to lend special plausibility to this fear: the widespread use of the bacterium *E. coli*, whose normal habitat is the human intestine, and the fact that plasmids which confer resistance against certain antibiotics are often employed to ferry foreign pieces of DNA into microbes. One conjecture was that the foreign DNA might turn a normally benign *E. coli* bacterium into a form which produces a dangerous disease; this bacterium could then infect someone, probably a laboratory researcher, and be spread throughout the population; finally, since the human body is quite accustomed to the presence of *E. coli*, it might not fight back against this dangerous variant, and the fact that the bacterium is resistant to one or more antibiotics would make medical treatment that much more difficult.

Most of the set pieces of the genetic engineering debate took place in the US where this research was generally furthest advanced. Two landmarks were the international conferences held in New Hampshire in June 1973 and the Asilomar Center, California, in February 1975. These discussions led eventually to the drafting of guidelines for genetic engineering research in countries which had the scientific expertise to carry it out, and produced a ban on certain types of experiment. It was not permissible, for example, to try to insert the gene which codes for cholera toxin into other bacteria that can live in the human gut.

The first set of guidelines laid down by bodies such as the US National Institutes of Health and Britain's Genetic Manipulation Advisory Group were quite stringent. They defined the laboratory conditions under which particular types of experiments could be attempted. Most importantly, they specified the precautions that must be taken to prevent microbes escaping, and what kind of microbes could be employed. By 1983 many of the restraints on genetic engineers had been loosened, although a considerable number of important safeguards remain. To understand the scientific basis for this shift towards a liberalization of the rules governing genetic engineering, it is necessary to be specific about what putative dangers should be considered, and the steps that can be taken to prevent them.

Two crucial points must be made at the outset. First, no-one can argue with complete honesty that there is not, and never will be, some conceivable risk in genetic engineering. The concept of zero-risk is virtually meaningless in any area of human activity. Even the simple act of breathing entails the risk of inhaling germs and contracting a disease. The only responsible attitude to any new technology is to find out as much as possible about its conceivable disadvantages and place them in the context of everyday risks. Any extra risk can then be weighed against known or supposed benefits.

Second, discussions concerning the safety of genetic engineering centre on the chances of accidents occurring. There is little doubt that someone sufficiently skilled and deranged could *deliberately* construct a new microbe that would pose a threat to human health. It is, however, by no means clear that any such microbe would be substantially more dangerous than existing, natural agents such as encephalitis viruses and the bacteria responsible for botulism. This is clearly a separate problem from that of the regulation of genetic engineering in universities and industry, and it is difficult to see how a ban on future genetic engineering in universities and industry would help stop scientific knowledge being perverted for destructive ends.

Thus, the questions that biotechnologists must answer concern the possibility of genetic engineering accidents and their effects. The vast majority of biologists now concur that the risks are negligible for several reasons. These stem from the fact that a pathogenic (disease-producing) organism must possess several distinct characteristics, and any creation by genetic engineers will be safe unless all are present. Most notably the organism must be able to survive in the special environment of the human body (assuming the particular case of human diseases); it must be capable of being transferred from one person to another; and it must somehow produce symptoms of a disease, perhaps by manufacturing some toxic substance.

The genetic engineers' first line of precaution is to prevent microbes escaping from the laboratory. This can include the containment of organisms in air-tight chambers and the thorough sterilization of all equipment before and after use. Microbiologists have had years of experience in handling very dangerous natural organisms – smallpox viruses, cholera bacteria and so on. There have been few accidents and no known cases in which these have spread disease to the general public.

Despite this good safety record, genetic engineering does not rely on these measures alone. The most important safeguards are provided by what is known as biological containment. Much of the early concern centred on the fact that most genetic engineering experiments involved *E. coli*. This implied that if a bacterium escaped from the laboratory it would have a good chance of becoming established inside the human body. However, the strain most often employed, *E. coli* K–12, is a feeble creature. This is not surprising since these bacteria have been grown for many generations in the luxuriant surroundings provided for them in laboratories. The result of this pampering has been to produce bacteria which would find it very difficult to survive if they happened to find their way out of this comfortable environment and into the much harsher conditions prevalent in the human body. *E. coli* K–12 is thus biologically contained, as without an artificial environment it cannot thrive.

Even this degree of enfeeblement in microbes for genetic engineering was considered insufficient to provide a good margin of safety. Microbiologists then set about breeding strains of *E. coli* which need to be provided with unusual chemicals if they are to live – chemicals that are not found in the bodies of human beings, other animals or plants. One such strain, developed by Roy Curtiss III in Alabama, is *E. coli* X1776, named in honour of the American bicentennial. This requires several unusual chemicals to live and is sensitive to bile salts (found in the intestine), antibiotics and detergents. Curtiss's bacterium has been approved for use in those genetic engineering experiments which specify that the microbe employed must be at least 100 million times less able to survive in nature than normal E. coli K–12.

A critical factor in initiating the recombinant DNA debate was a proposal put forward in the early 1970s to insert, into E. coli, genetic material from SV 40, a monkey virus. SV 40 is known to cause cancer in mice and the idea of introducing its genes into E. coli clearly required the most careful thought. This prompted a series of experiments with another virus, the results of which had a profound influence on discussions about risks. Polyoma virus is the most infective tumour virus known for hamsters, and it was decided to insert this virus into E. coli and inject it into animals. The fact that no tumours were caused indicated that once inside the bacterium this virus was much less dangerous than normal.

The 1982 guidelines which apply to all work carried out in connection with the US National Institutes of Health – the major source of funding in that country – prohibits six main classes of experiments, including the cloning of DNA from certain disease-causing organisms, the cloning of genes which code for toxins that harm vertebrate animals and the transference of antibiotic-resistance genes into organisms that cause diseases of humans, animals or plants.

The climate of opinion has now shifted dramatically towards an acceptance that genetic engineering – especially the kind that biotechnologists want to perform – poses no serious hazard. Indicative of this new attitude was the 1983 decision of the US Recombinant DNA Advisory Committee to allow scientists to carry out field tests involving genetically engineered plants and microbes. One such test involves growing tomato and tobacco plants which carry genes from bacteria and yeasts, and another will release genetically engineered microbes whose natural counterparts cause damage to the leaves and flowers of crops. The microbes found in the wild act as 'ice nucleation' centres – that is, they facilitate the formation of ice crystals on plants. The genetically altered microbes have been modified with the aim of removing this ability to promote ice crystallization.

This decision, and the certainty that other new proposals will be put forward by genetic engineers that require careful consideration, makes it unlikely that the present regulations will be swept away. Some scientists fiercely resent government-imposed guidelines, which they see as interfering with their 'traditional right' of free enquiry. Some have even compared their situation with the 'gagging' of Galileo, a claim which it is difficult to take seriously. The shock of finding themselves caught in the glare of public scrutiny has profoundly altered the way many biologists view their work and its place in society. Some bitterly regret that their action in drawing attention to the *conceivable* dangers of genetic engineering (as they were seen in the mid-seventies) were interpreted as a warning of imminent disaster. Others believe that it was to their credit that an awareness of their responsibilities enabled the whole matter to be discussed fully, thereby preventing a headlong plunge into unbridled experimentation.

Questions about genetic engineering will undoubtedly continue to be raised from time to time. However, current research, carried out under the regulations, carries very little risk, and its potential benefits amply justify the continuation of genetic engineering as a tool for biotechnologists. In this brief discussion it has only been possible to outline some of the main themes of a complex and often highly technical topic. The Bibliography lists some of the many publications that have examined the safety of genetic engineering from almost every perspective.

## Where will the bioriches flow?

In the eighteenth century, the development of the steam engine launched Britain on its path towards domination of the world's economy, while Japan's recent meteoric rise in global trade has been boosted by the sophisticated use of microelectronics. Similarly, the innovations of mass production greatly strengthened the US's power, and Germany reaped immense wealth from the growth of its chemical industry. The economic advantages of being among the first to capitalize on new technologies (not necessarily to invent them) are enormous – which nations will make the most of the bioindustrial revolution? The US and Japan – not necessarily in that order – are in the best position right now, with the US having a clear, but not yet commanding, lead in genetic engineering, and Japan having an edge in the kind of technology needed for large-scale fermentations, with about 80 per cent of the patents in this area. But it is certainly not too late for Britain, France, Switzerland, West Germany, Denmark and several other developed countries to seize a large slice of the cake. Furthermore, there is little reason why, given the necessary commitment, a number of developing countries should not become more involved in the lower-technology, less capital intensive areas of bio-technology.

The pattern of investment differs markedly between the US and Japan. In the former, especially in California and Massachusetts, dozens of small firms have been established in the last ten years. Many have considerable expertise, and are receiving massive financial backing from multinational corporations as diverse as Standard Oil, Dow Chemicals, International Nickel, General Foods and Bendix. Huge amounts of money are also being funnelled into basic research in universities and other public institutions, often in return for an option on any patent licences that may result from the investigations. For example, the German chemical giant Hoechst plans to plough $50 million into Massachusetts General Hospital's molecular biology laboratory.

Meanwhile in Japan, the government is instrumental in promoting biotechnology. The Ministry of International Trade and Industry is allocating $110 million over the next ten years to supplement the expertise of industrial firms in large and sophisticated fermentation processes. Their prowess in biotechnology is amply demonstrated by a virtual monopoly in world trade of some amino acids, enzymes and food additives, and Japan has more industrial experience in working with immobilized enzymes and cells than any other nation. Government funds of about $20 million in 1981 aided industry to increase its knowledge of monoclonal antibodies and genetic engineering, including the production of interferon. Japan already earns over $50 billion a year by exploiting microbes – about 5 per cent of its gross domestic product.

In Britain, the Spinks report, *Biotechnology*, published in 1980, called for a major commitment to biotechnology, and the Government's response was a skimpy White Paper which offered only bland platitudes and paid little more than lip-service to the idea that biotechnology could be the formation of vast and profitable industries within a few years. But more recently, the message seems to have struck home with a grant of £16 million from the Department of Energy to eight research institutions. Equally encouraging is the decision by the University Grants Committee, at a time when budgets are being slashed in virtually all areas, to earmark £2.4 million over 1983–5 for the development of biotechnology.

Recent estimates made by the EEC indicate that the UK is spending $46 million of public money on research and development centred on biotechnology. Comparable figures from other areas include West Germany ($36 million), France ($31 million), the EEC as a whole ($146 million), US ($200 million) and Japan ($50 million). However, when spending on wider research related to biotechnology is taken into account, the balance shifts. While the UK figure rises only to $59 million, West Germany's goes up to $132 million, France's to $84 million, the entire EEC's to $355 million and the US's to $550 million.

Of more than twenty firms set up in Britain to exploit new areas of biotechnology, the best-established is Celltech. This firm was created by the National Enterprise Board in 1980, and its financial backers include the Prudential Insurance Co. and the Midland Bank. Celltech has close links with the Medical Research Council which has some of the most prestigious laboratories in the world. This link has brought Celltech its first marketable product, an antibody which latches on to interferon molecules.

British industry is investing in biotechnology both through specialist research centres associated with universities and through the establishment of private research facilities. The breadth of interest in biotechnology is exemplified by the support being given to the new Biocentre at Leicester, which is receiving money from brewers (Whitbread), tobacco manufacturers (Gallaher), food processors

(Dalgety-Spillers) and engineers (John Brown). Research and development is also expanding rapidly within industrial companies, ICI, Shell, Glaxo, Burroughs Wellcome, G. D. Searle, and Tate and Lyle being just a few of the massive companies to take a stake in biotechnology.

Britain's prosperity will certainly hinge on its ability to establish the industries that will underpin the world economy during the next century. The opportunities of the microchip were largely scorned and the chance to lead in that industrial revolution wasted, but there is no reason why the pattern should be repeated in biotechnology, so long as the political will and industrial imagination can be mustered.

# Biotechnology and the Third World

The population of our planet is divided in many ways, but the most profound differences are to be found in health and access to effective medical treatment. Put bluntly, the inhabitants of the Third World die younger, in more pain and of different diseases than those in affluent countries. In much of Asia, Africa, and Central and South America, infectious and parasitic diseases are still rampant. Millions are struck down each year by cholera, malaria, sleeping sickness and a hideous array of debilitating and fatal diseases. Biotechnology, as we have seen, *could* do much to lift the scourge of these diseases, but will the necessary commitment be forthcoming? If new vaccines and therapies are to be developed, a much higher priority must be assigned to these areas. The US Institute of Medicine calculates that cancer research in that country is funded at the rate of $209 a year for each case of the disease and cardiovascular disease receives $8 per case. The equivalent figures for schistosomiasis and malaria are 4.5 and 2 cents.

The development of a new drug can take years and cost millions of pounds. Once a substance has shown promise in laboratory experiments, a decision must be made on whether to initiate the work that consumes most time and money – testing its effects on animals and, eventually, human volunteers. Even the largest pharmaceutical companies can only follow up a few of the leads provided by their research scientists. Among the many criteria applied when taking the decision to continue development work on a potential new drug is one that weighs very heavily against the Third World – the profit potential of the product. There is little incentive to invest in a drug which will find its major markets in countries with hardly any money. Since most pharmaceutical firms are under a legal obligation to act in the interest of their shareholders, it is unlikely that this situation will change.

Thus, if the Third World is to benefit fully from the opportunities that biotechnology presents for an attack on parasitic and infectious diseases, alternative funding must be found. The World Health Organization and other international institutions have research laboratories, but there is an urgent need for new non-profit-making development and testing facilities as well as for increased effort in basic research.

The United Nations Industrial Development Organization (UNIDO) has begun planning an international biotechnology research centre which will focus particularly on the problems of the developing countries. In the medical area, the primary need is to follow up more of the many discoveries that have been made in existing research centres and universities, and develop them into practical processes that could transform the quality and length of life for millions. Although under the present system pharmaceutical companies could not justify funding development work themselves, they could be encouraged to hand over information on projects which they view as commercially unsound but of possible medical value. This is not as unrealistic a proposition as it may at first appear. Pharmaceutical companies are acutely conscious of their public image, and a chance to be seen to aid the Third World, without any significant extra costs, could prove attractive.

Another factor which hinders the development of many types of drugs is the enormous investment in testing which is demanded by governments. Regulations take little account of the nature and extent of the illness concerned when stipulating the required safety and efficacy tests. It is clear that any drug which is intended to treat a relatively minor illness should be subject to the most intense scrutiny before it is licensed for general use, especially if other treatments are already available. When, however, we are considering common and fatal diseases – and not just those that primarily afflict the developing nations – the balance of risks versus benefits is very different. If governmental controls are such that no commercial company finds the research and development worthwhile, we shall clearly avoid the danger of unsafe drugs – there will not be any new drugs.

Any relaxation of the rules governing the commercial pharmaceutical industry is bound to be politically unacceptable, especially if it appeared that 'second-class' safety testing was considered adequate for some types of patient. Yet an international, non-profit-making research and production organization might be able to accelerate the introduction of some new drugs. It would be seen that medical benefits and risks were the main factors being taken into account and that considerations of sales income played no part in the decision.

It is important to note when discussing the 'safety' of drugs, that this is a relative term. No drug is absolutely safe in the sense that it will never produce some sort of undesirable side-effect in some patients. The public is certainly right to demand that any drugs put on the market should be tested as required by government regulations and any adverse effects which subsequently appear should be promptly and fully investigated. It is not realistic however to insist that all side-effects be eliminated – unless, that is, we are prepared to halt all work on new drugs and withdraw all the existing ones.

The US Office of Technology Assessment estimates that by the end of the century biotechnology could be supplying 20 per cent of the nation's energy. If this target is achieved, fuel alcohol will play a major role, probably the dominant one. The main raw materials consumed during the manufacture of fuel alcohol – sugar cane and wheat in particular – are currently used as food. Does more fuel for the rich mean less food for the poor? At least until efficient methods can be designed which employ wood and other cellulose-rich plant materials to produce fuel alcohol, this will remain a very valid question. The Kenyan government withdrew its support for fuel alcohol programmes in 1982, when it discovered that food imports had soared to replace the agricultural produce diverted to supply the new industry. Countries should be wary of tailoring their agriculture towards crops which can yield fuel alcohol. Unfortunately, the temptation is especially great for developing countries which desperately need foreign currency.

Indeed, one of the major global economic problems that will be thrown up by the bioindustrial revolution concerns sugar cane – a major prop in the economies of many countries, especially in the Caribbean. Increased use of fuel alcohol should push up the demand for sugar cane, but the growth in the market for alternative sweeteners, including fructose and aspartame, will depress prices. The overall impact of these opposing forces can only be guessed at, but in the short term the already shaky economies of sugar-producing nations may well be further undermined.

The interaction of other areas of biotechnology are even more difficult to predict. Economic advisers in every country, especially in the Third World, will need to consider the likely effects on, for example, their energy needs. Methane generated from wastes should help cut oil imports and, in the longer term, the production of hydrogen from water will have the same effect. Yet a decreased death rate due to new medicines will increase the population and its energy demands. The extra people will also need to be fed. Diverting crops into the production of fuel alcohol will exacerbate food shortages, but the possibility of new crops grown on previously unusable land will have the opposite effect. Single-cell protein factories will consume plant starches, but yield more nutritious foods for domestic animals, or preferably, for humans.

## The scientist as entrepreneur

No sooner were biologists recovering from the shock of the genetic engineering debate than they were engulfed in another controversy – the relationship between universities and industry. Little of the vast commercial potential of the newer biotechnologies would have been possible without the basic work undertaken by universities and other publicly funded research institutions. Until recently, virtually every significant development in genetic engineering and the production of monoclonal antibodies was supported by tax-payers or charities. Who is likely to reap the financial benefits that will accrue from the commercialization of this new knowledge? The answer in many cases is becoming increasingly clear – a rapidly growing band of top-rank scientists is leaving the universities to set up commercial enterprises whose main assets are the knowledge and skills of their researchers, knowledge and skills which were largely gained during years of publicly funded work. Even more controversy has been stirred up by the fact that some biologists have retained their university posts at the same time as sitting on the boards of such companies.

This trend has become particularly pronounced in the US, where most of the new biotechnology firms are staffed at the top scientific levels with people who only a year or so ago were heading major laboratories in the nation's finest universities. The desirability of various solutions to this problem, indeed the very perception that there is a problem at all, depends largely on the political stance of the observer.

Those in favour of the present system by which scientists move from academia to industry can put forward a convincing case. Biotechnology is the application of biology within an industrial context, and they argue that if there is no movement of scientists from basic research into applied development in commercial operations then the full potential of biotechnology will not be realized. Certainly there is little reason to believe that the universities are either willing or able to develop and market products which arise out of investigations within their laboratories. They also argue that scientists should have the same rights as everyone else to sell their labour (or knowledge) as best they can on an open market. Furthermore, if the interchange of scientists between academic and industrial life stimulates the growth of new industries based on biotechnology, everyone will gain through better medicine, more food, greater energy supplies and all the other benefits on offer. Finally, they note that, although these questions are new to biologists as a group, other types of scientists – chemists, for example – have for many years had close links with major industries and that there is no evidence that these arrangements have generally had adverse consequences.

The issues raised by those who believe that the present system is too much of a free-for-all are usually more complex, but are nonetheless very important. At one extreme – unlikely ever to be reached in practice – most of the best biotechnologists might be lured away to industry where commercial exigencies would ensure that they devoted their efforts to high-profit programmes. These may often not be the most socially useful (however defined) applications of biotechnology. Certainly one could expect that, within the pharmaceutical area, more effort would be devoted to the quest for cold cures than therapies for leprosy. The profit motive can have more surprising implications. The US Office of Technology

Assessment has noted, for example, that 'until recently, the US commercial seed companies, with one or two exceptions, have not been interested in wheat-breeding programs as a profit making venture'. The reason for this is that a farmer need buy only one batch of seeds and then retain some of the harvested seeds for replanting next year; one-off sales are bad for business. Therefore the development of higher-yielding wheat varieties had to come mainly through publicly funded research. Complete neglect of economically valuable, but unprofitable (in the narrow sense) research is a real danger, if the growth of biotechnology is almost entirely in the hands of commercial enterprises.

Within the universities the most active, and often acrimonious, debate concerns the distortion of traditional academic values of shared information and freedom of enquiry. These questions lie at the heart of scientific discovery – the very process that has powered the upsurge of biotechnology.

The tensions created by university scientists also being deeply involved in commercial enterprises was seen at its most acute in several US universities. The close links between Genentech, a genetic engineering firm, and staff at the Department of Biochemistry and Biophysics at the University of California in San Francisco caused much dissension. An investigating committee noted that some scientists they interviewed 'believed that the manner in which the particular contract [with Genentech] was carried out led to serious disruption within the department. A recurrent theme was that people were loath to ask questions or give suggestions in seminars or across the bench for there was a feeling that someone might take an idea and patent it, or that an individual's idea might be taken to make money for someone else'. They also foresaw greater problems when more than one firm had intimate links with researchers in a department. Subsequently, members of that department were serving as directors of three competing companies.

The souring of relationships as a result of such ties has been a pernicious influence in recent years. In the US, particularly, universities are searching almost frantically for solutions which will enable them to retain both their scientists and their reputations. Complex deals are constantly being proposed with the aim of gaining industrial funding, making sure that universities benefit from patents arising from their work, and relieving staff of the pressure to work on projects simply because they look profitable in the short term. The conflict between the traditional values of university and industrial research is stark indeed. On the one hand there exist ideals of open communication of information and the freedom to pursue studies for their intrinsic interest; on the other we find the need to guard secrets and to devote precious time and money only to investigations which promise to be profitable.

# Genetic Eldorado?

In October 1980 shares in Genentech, a company with only a handful of employees and no products, were offered to the public. Wall Street was so smitten with the biotechnology bug that within twenty minutes of the start of dealing the share price for this apparently unprepossessing company had rocketed from $35 to $89. At the peak of the stock market's flirtation with biotechnology, the mere hint of a cloned gene was enough to send investors scurrying to their stockbrokers, even when neither of them knew the commercial viability of this latest wonder of science. Subsequent share offers were greeted with only slightly less exuberance as other biotechnology firms took the plunge into the stock market in the US. In Britain, the demand for shares in Amersham International, a company that supplies many of the chemicals used by genetic engineers, greatly exceeded the expectations of the Government's financial advisers when it was decided to invite private investment in this hitherto publicly owned firm.

The rush to take a stake in biotechnology made multimillionaires of several scientists who had set up small, high-technology companies. Predictably, there has been a backlash among investors as they began to realize that fortunes in biotechnological products were not to be made overnight. Share prices dropped back, but still remain at remarkably high levels when compared with major companies with established profit records over many years. Most people involved with biotechnology view this retreat as a desirable development; if the biobubble had become even further inflated its eventual, and inevitable, burst could have proved disastrous for the industry, destroying valuable and necessary confidence in its future.

Dozens of small biotechnology firms are in the race to produce genetically engineered microbes that will manufacture highly profitable proteins, and some are bound to collapse over the next few years. Many of the successful ones are likely to be taken over by today's corporate giants, once they have solved the basic technical problems of producing new materials and require large infusions of capital to set up major manufacturing installations. The implications of this for employment are difficult to foresee.

The bioindustrial revolution is creating a huge demand for scientists and engineers who are trained in genetic engineering, fermentation technology, microbiology and numerous other specialities. In some areas the shortage of skilled personnel is already acute, and those with the right qualifications can command salaries far in excess of those thought possible only a decade ago. Governments and universities throughout the developed world are urgently seeking to expand high-level education in the subjects which underpin biotechnology. In Britain alone, nearly a dozen new courses at universities and colleges are being established and more are sure to follow. Thus, the employment prospects for those who are able to pursue these careers are bright.

The effects of the growth of biotechnology on employment as a whole are far less clear and no comprehensive estimates have been produced. The US Office of Technology Assessment suggests, for example, that 30,000 to 75,000 workers might be needed to produce $14.6 billion worth of chemicals from genetically engineered microbes. Since the products they considered are already being manufactured by the chemical factories, there would clearly be job losses in the more traditional industries. A rough estimate indicates that there would be little net gain or loss in employment caused by a shift towards biotechnology in this area.

The Organization for Economic Cooperation and Development (OECD) took a similar line in its 1982 report, stating that the growth of biotechnology will not substantially affect total employment in the short term. Looking further into the future, however, it is possible to foresee more jobs being created, as biotechnology yields entirely novel products rather than replacements, direct or indirect, for existing products and manufacturing methods.

## Towards a more stable world economy?

Since the industrial revolution, the global economy has been characterized by ever-increasing demands for energy and metals. Until now the demand for energy has been met chiefly by exploiting fossil fuels, while metals have been supplied by increasingly sophisticated mining techniques. We all know that there are limits on how long we can continue to pursue this path, even though we act most of the time as if no such limits existed. When people come to look back on the bioindustrial revolution of the late twentieth and early twenty-first centuries they will, perhaps, see its greatest contribution as being the transformation of the society from one dependent on non-renewable materials, and hence inherently unstable, into one based largely on a power source which will be with us for millions of years to come – the sun.

In the next century, plastics derived from living organisms may perform many of the jobs now done by metals. Since all organisms are, in effect, powered by the sun's energy, we can expect a virtually unlimited supply of these materials. Any remaining need for metals could be met by microbial mining and, more importantly, by using microbes to recycle the metals that have already been extracted from the Earth by conventional mining methods.

Our energy needs could similarly be met by sunlight with plants being used to manufacture fuel alcohol, and methane being generated from wastes. If reliable hydrogen generators can be made to work with an efficiency of 10 per cent – that is, converting 10 per cent of light energy into energy stored in hydrogen gas – an area of only 500,000 sq km (193,000 sq miles) of such generators would be able to supply all of the world's current energy needs. This is only about 7 per cent of the area of Australia.

These are grandiose ideas, but no-one who has witnessed the almost incredible pace at which biotechnology has moved in the last decade would feel confident in dismissing them as mere pipe-dreams.

# Bibliography

These publications are recommended as sources of further information and opinion.

A.T. Bull, G. Holt and M.D. Lilly  *Biotechnology: International Trends and Perspectives*  Organisation for Economic Co-operation and Development, 1982 – considers economic, political and social impacts of biotechnology

J. Cherfas  *Man-made Life*  Basil Blackwell, Oxford, 1982 – a detailed, but lucid, explanation of the principles and applications of genetic engineering

D.A. Jackson and S.P. Stich (eds.)  *The Recombinant DNA Debate*  Prentice-Hall, Englewood Cliffs, New Jersey, 1979 – presents several opposing views on the ethics and safety of genetic engineering

S. McAuliffe and K. McAuliffe  *Life for Sale*  Coward, McCann & Geoghegan, New York, 1981 – a popular account of genetic engineering

J.R. Norris and M.H. Richmond (eds.)  *Essays in Applied Microbiology*  John Wiley, Chichester, 1981 – eleven interesting essays on the practicalities of biotechnology

The Office of Technology Assessment  *Genetic Technology: A New Frontier*  Westview Press, Boulder, Colorado/Croom Helm, London, 1982 – a comprehensive and clear report prepared for the US Congress. Especially useful for information concerning potential markets for biotechnology products

*Scientific American*  W.H. Freeman, San Francisco, September 1981 – a special issue which highlights several important applications of biotechnology

A. Spinks (chairman)  *Biotechnology: Report of a Joint Working Party* – senior British scientists outline actions necessary to take advantage of biotechnology

L. Stryer  *Biochemistry* (2nd edition)  W.H. Freeman, San Francisco, 1981 – an excellent textbook

J. Watson and J. Tooze  *The DNA Story*  W.H. Freeman, San Francisco and Oxford, 1981 – a documentary history of the recombinant DNA debate

J. Watson, J. Tooze and D.T. Kurtz  *Recombinant DNA: a Short Course*  W.H. Freeman, New York and Oxford, 1983 – a clear and concise explanation of the principles of genetic engineering

A. Wiseman (ed.)  *Principles of Biotechnology*  Surrey University Press, 1983 – covers several major themes of biotechnology

E. Yoxen  *The Gene Business*  Pan, London, 1983 – examines the social and political implications of the rise of biotechnology, as well as the technology itself

The following periodicals report on current developments in biotechnology:

*Trends in Biotechnology*  Elsevier Science Publishers, Cambridge – provides much technical information on the principles and practicalities of biotechnology

*Bio/Technology*  Nature Press, New York – news and features on the biotechnology scene, especially in the USA

Finally, three weekly journals provide a good coverage of science and its implications for biotechnology:

*Nature*  Macmillan, London

*Science*  American Association for the Advancement of Science, Washington

*New Scientist*  IPC Magazines, London

# Index

186